Maritza Jaramillo

Malaria Immunopathology

Maritza Jaramillo

Malaria Immunopathology

Signaling and Cellular Mechanisms Involved in Hemozoin-inducible Proinflammatory Events

VDM Verlag Dr. Müller

Impressum/Imprint (nur für Deutschland/ only for Germany)
Bibliografische Information der Deutschen Nationalbibliothek: Die Deutsche Nationalbibliothek verzeichnet diese Publikation in der Deutschen Nationalbibliografie; detaillierte bibliografische Daten sind im Internet über http://dnb.d-nb.de abrufbar.

Coverbild: www.purestockx.com

Verlag: VDM Verlag Dr. Müller Aktiengesellschaft & Co. KG
Dudweiler Landstr. 99, 66123 Saarbrücken, Deutschland
Telefon +49 681 9100-698, Telefax +49 681 9100-988, Email: info@vdm-verlag.de
Zugl.: Quebec, Universite Laval, 2005

Herstellung in Deutschland:
Schaltungsdienst Lange o.H.G., Berlin
Books on Demand GmbH, Norderstedt
Reha GmbH, Saarbrücken
Amazon Distribution GmbH, Leipzig
ISBN: 978-3-639-13633-3

Imprint (only for USA, GB)
Bibliographic information published by the Deutsche Nationalbibliothek: The Deutsche Nationalbibliothek lists this publication in the Deutsche Nationalbibliografie; detailed bibliographic data are available in the Internet at http://dnb.d-nb.de.

Cover image: www.purestockx.com

Publisher:
VDM Verlag Dr. Müller Aktiengesellschaft & Co. KG
Dudweiler Landstr. 99, 66123 Saarbrücken, Germany
Phone +49 681 9100-698, Fax +49 681 9100-988, Email: info@vdm-publishing.com
Quebec, Universite Laval, 2005

Printed in the U.S.A.
Printed in the U.K. by (see last page)
ISBN: 978-3-639-13633-3

TABLE OF CONTENTS

ACKNOWLEDGEMENTS

I would like to give special thanks to my PhD supervisor and friend Dr. Martin Olivier. First, I want to express him my gratefulness for having given me the opportunity to join his team. I will always remember his important contribution to my presence in the scientific community and in this wonderful country, Canada. During the past years in Dr Olivier's laboratory, I have had the chance to learn from his vast expertise in the Immunology and Microbiology fields, but most important from his creative mind and his passion for science.

I want to thank to all the former and current members of Dr. Olivier's team and to those who shared our working environment. During my Graduate studies in Quebec and Montreal I had the chance to meet and learn from many people who became not only my colleagues but also my friends. Cecilia Maggie, Adelaida, Julien, Marianne, and Marc, I will never forget your support in the difficult times as well as the beautiful moments that we shared.

At last but not the least, I want to thank my family, which is the most precious gift I have. Since this is a great opportunity to express my gratitude to them, I prefer to do this in my own language. Ustedes, mi familia, saben mas que nadie cuan duro has sido este camino pues lo han vivido conmigo a paso a paso desde el comienzo. Por eso quiero agradecerles desde el fondo de mi corazón por su presencia constante, su paciencia y sobretodo su amor para comprender y respetar mis decisiones y para apoyarme en mis proyectos. Sé que han compartido todas mis alegrías pero también todas mis tristezas y mis luchas, asi que en la culminación de esta etapa quiero decirles con todo mi amor y gratitud que lo que soy y lo que he logrado se los debo a ustedes que son mi fuerza y mi estímulo en este paso por la vida.

ABBREVIATIONS

AA	Arachidonic acid
Ab	Antibody
AD	Actinomycin D
AP-1	Activator protein-1
APC	Antigen-presenting cell
ARE	Adenylate/urydilate-rich elements
AUBF	Adenosine-uridine-binding factor
C/EBP	CCAAT/enhancer binding protein
cAMP	Cyclic adenosine 5' -monophosphate
CAPE	Caffeic acid phenethyl ester
CAT	Cationic amino acid transporter proteins
CBP	CREB-binding protein
CCL	CC chemokine ligand
CCR	CC chemokine receptor
CM	Cerebral malaria
CMV	Cytomegalovirus
CQ	Chloroquine
CRE(B)	Cyclic AMP-response element-(binding protein)
CXCL	CXC chemokine ligand
CXCR	CXC chemokine receptor
DC	Dendritic cell
DFX	Desferrioxamine
DV	Digestive vacuole
ECM	Experimental CM
EM	Electron microscopy
EMSA	Electrophoretic mobility shift assay
ERK	Extracellular signal-regulated kinase
GAPDH	Glyceraldehyde-3-phosphate dehydrogenase
GAS	Gamma activated site
GPI	Glycosylphosphatidylinositol
GSH	Reduced glutathione
H. columbae	*Hemoproteus columbae*
H_2O_2	Hydrogen peroxide
Hb	Hemoglobin
HETE	Hydroxyeicosatetraenoic acid
HIV-1	Human immunodeficiency virus type 1
HNE	4-hydroxynonenal
HO	Heme oxygenase
HRP	Histidine-rich protein
HTLV-1	Human T lymphotropic virus type 1
HUVEC	Human vascular endothelial cells
HZ	Hemozoin
IκBα	Inhibitory protein κBα
i.v.	Intravenous(ly)
ICAM	Intracellular adhesion molecule
IFN	Interferon
IKK	IκB kinase
IL	Interleukin
iNOS	Inducible nitric oxide synthase
IP-10	IFN-γ-inducible protein -10
iRBC	Infected red blood cell
IRF-1	IFN-γ responsive factor-1
ISRE	IFN-stimulated response element

IVBMC	Intervillous blood mononuclear cells
Jak	Janus kinase
JNK	c-Jun amino-terminal kinase
L. donovani	*Leishmania donovani*
L. major	*Leishmania major*
LPS	Lipopolysaccharide
Mϕ	Macrophage
MAPK	Mitogen-activated protein kinase
mCK	Mouse chemokine
MCP	Monocyte chemoattractant protein
mCR	Mouse chemokine receptor
MEK1/2	MAPK kinase 1/2
MEKK1	MAPK/ERK kinase kinase-1
MIF	Migration inhibitory factor
MIP	Macrophage inflammatory protein
MRP	Myeloid-related protein
MSU	Monosodium urate
MyD88	Myeloid differentiation factor 88
Nϕ	Neutrophil
NAI	Natural acquired immunity
NEMO	NF-κB essential modulator
NF-κB	Nuclear factor κB
NIK	NF-κB-inducing kinase
NO	Nitric oxide
NOX	NADPH oxydase
O_2^-	Superoxide anion
P. falciparum - Pf	*Plasmodium falciparum*
PBMC	Peripheral blood mononuclear cells
PG	Prostaglandin
PI3K	Phosphatidylinositol 3-kinase
PKC	Protein kinase C
PM	Placental malaria
PMA	Phorbol ester
Poly B	Polymixin B
PP	Protoporphyrin
PTK	Phosphotyrosine kinase
PTP	Phosphotyrosine phosphatase
PV	Parasitophorous vacuole
PVM	PV membrane
RBC	Red blood cell
Redox	Reduction-oxidation
RNI	Reactive nitrogen intermediates
ROS	Reactive oxygen species
RPA	RNase protection assay
S. mansoni	*Schistosoma mansoni*
sHZ	Synthetic hemozoin or β-hematin
SOD	Superoxide dismutase
STAT	Signal transducers and activators of transcription
T. cruzi	*Trypanosoma cruzi*
T. gondii	*Toxoplasma gondii*
TLR	Toll-like receptor
TNF	Tumor necrosis factor
UTR	Untranslated region
VSMC	Vascular smooth muscle cells

CHAPTER 1
MALARIA

PARASITE CLASSIFICATION

The genus *Plasmodium* is classified in the order *Haemosporida*, class *Hematozoa*, phylum *Apicomplexa*. The apicomplexans are all parasites, characterized by the "apical complex", a distinctive structure that inspired the phylum name (1). The *Haemosporida* are digenetic parasites; this is maturation of gametes and fertilization and sporogony occur in the hematophagous invertebrate vector while the rest of the life cycle is completed in the blood of the vertebrate host. The transmission by arthropod vectors and the origin of complex life cycles would have been a parasite adaptation that evolved because it favors its rate of transmission (2).

The genus *Plasmodium* consists of nearly 200 known species that parasitize reptiles, birds, and mammals, including rodents and primates. Four species are parasitic to humans: *P. malariae*, *P. ovale*, *P. vivax*, and *P. falciparum*, the agent of malignant tertian malaria, the most severe of human malarias (3). To account for the virulence of *P. falciparum* it has been postulated that it is a recent human parasite, acquired by a host switch, perhaps from domestic birds (4). Genetic studies have revealed that *P. falciparum* is phylogenetically closer to the *Plasmodium* species parasitic to birds than to the other species that parasitize humans and other primates (5). However, the time of divergence between *P. falciparum* and the avian *Plasmodium* species estimated from their phylogeny is in the tens of millions, and thus is inconsistent with a direct lateral transmission (*e.g.*, new associations of parasite-vertebrate host) from birds to humans (6).

PARASITE STRUCTURE AND LIFE CYCLE

Plasmodium species, as members of *Apicomplexa*, share many common morphological features. Each of the developmental stages in the life cycle of malaria parasites exhibits a remarkable conservation; however, structural differences exist, among reptilian, avian, and mammalian plasmodia (7). *Plasmodium* is a genomically complicated eukaryote, able to constantly change its gene expression to generate a sequence of forms that exploit most efficiently quite different environments: liver and RBC in humans; gut, vascular system and salivary glands in the mosquito. Thus, structural, biochemical and molecular biological aspects are different among various stages of the complex life cycle of the parasite, which comprises a sexual stage in the mosquito and two asexual phases with multiplication in the vertebrate host (8).

Liver Stages

Plasmodium sporozoites are transmitted to the vertebrate host by the bite of infected female mosquitoes of the genus *Anopheles*. Mosquitoes inject parasites into the subcutaneous tissue, and less-frequently directly into the bloodstream. Within 60 min of inoculation, the sporozoites move to the liver. A series of complex molecular interactions between sporozoite proteins and hepatocyte surface molecules occurs during invasion and subsequent intrahepatic development (9). Over the course of 5-15 days depending on the *Plasmodium* species (*P. falciparum* ~5 days, *P. vivax* ~8 days, *P. ovale* ~9 days, *P. malariae* ~16 days), the invaded sporozoites undergo the process of asexual reproduction known as pre-erythrocytic or exoerythrocytic schizogeny. During the development of uninucleate sporozoites to mature pre-erythrocytic forms, several antigens have been identified, including the circumsporozoite protein (CSP) (10). *P. vivax* and *P. ovale* have a dormant stage, named hypnozoite, which remains in the liver from weeks to years before pre-erythrocytic schizogeny (11). This results in relapses of malaria infection; however, the biological basis of this stage remains unknown. In contrast, *P. falciparum* and *P. malariae* have no persistent phase and thus undergo pre-erythrocytic schizogeny. A pre-erythrocytic schizont contains 10,000 (*P. vivax*) to 30,000 (*P. falciparum*) merozoites that are released into the blood circulation and can invade only RBCs (10).

Erythrocytic Stages

The intrahepatic stages of the parasite are asymptomatic, but during the erythrocytic cycle the clinical manifestations of malaria occur. Pathogenesis depends on the parasite effects in the RBC population and in the immune response; an impact progressively amplified by repeated cycles of RBC invasion, intracellular growth, multiplication and re-invasion (Figure 1) (8). The duration of the erythrocytic cycle varies according to the *Plasmodium* species, being shorter in *P. falciparum* (36 to 48 h), *P. vivax* and *P. ovale* (48 h), and longer in *P. malariae* (72 h). The end of the cycle correlates with the periodic fevers and chills characteristic of malarial paroxisms, and their frequency gave rise to the names of tertian (*P. vivax, P. ovale*) and quartan malaria (*P. malariae*) for the different *Plasmodium* infections. Because of its fatal effects on the host, the infection caused by *P. falciparum* is known as malignant tertian malaria (12).

Extracellular merozoites are intrinsically short-lived; thus, within 1-2 min of release they select, adhere and invade a fresh RBC. The sequence of events has been analyzed most closely in *P. knowlesi*, but the available evidence of *P. falciparum* invasion suggests that the process is very similar in both species (8). Even though the RBC invasion takes only ~30 s, it is a very complex process, and is dependent on multiple interactions between specific receptors on the RBC membrane and ligands in the merozoite surface (reviewed in (13)). Release of parasite substances and internalization of RBC plasma membrane components lead to the formation of the membrane-lined parasitophorous vacuole (PV) into which the merozoite propels itself. Once inside the RBC, dense granule release from the parasite enlarges the PV membrane (PVM), and transition to ring stage occurs. During the ring stage, the parasite flattens into a thin biconcave disk, thicker around its perimeter where the elongated nucleus is present and thinner in the middle, giving it the appearance of a ring in Giemsa-stained blood smears (14). The parasite begins to feed on small aliquots of RBC hemoglobin (Hb) through a structure named cytostome, and to take up other nutrients transported in from the plasma. The ring stage enlarges and eventually grows into the more rounded trophozoite stage (8).

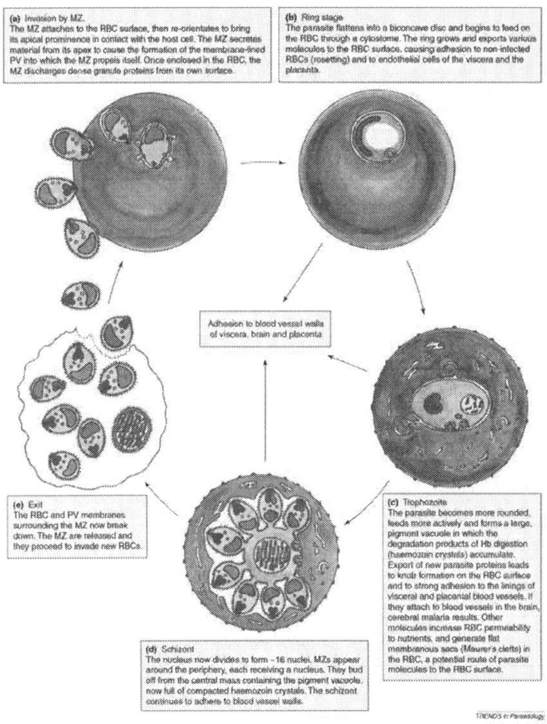

Figure 1. Intraerythrocytic Life Cycle of *P. falciparum*. Hb, hemoglobin; MZ, merozoite; PV, parasitophorous vacuole; RBC, red blood cell. Reprinted from *Trends in Parasitology,* Vol 19 (5):209-213, Bannister, L., and G. Mitchell, The ins, outs and roundabouts of malaria, Copyright 2003, with permission from Elsevier. Reference (8).

The trophozoite period is the most active in feeding, growth and RBC modification. The parasite avidly digests Hb, and heme produced during this process is crystallized into particles of dark-brown pigment known as hemozoin (HZ) or malaria pigment, which is stored within the digestive vacuole (DV) (15). New molecules are exported into the RBC leading to knob formation on the RBC surface, which cause its adhesion to non-infected RBC (rosetting) and to the vascular endothelium (cytoadherence) of the major organs in the body (brain, heart, lungs and submucosa of the small intestine) (16). Sequestration of iRBC in the microvasculature is a central feature of *P. falciparum* infection and is considered one of its main virulence factors, since not only it reduces parasite removal from the blood stream via the spleen, but it also leads to microvascular flow disturbance, increases vascular permeability and causes tissue damage. When this process occurs in the brain, it results in cerebral malaria (CM) (17).

The schizont is the erythrocytic phase in which the parasite undergoes a series of nuclear divisions by multiple fission, ~8 to 24 nuclei are generated (*P. falciparum* 8-24, *P. vivax* 12-24, *P. ovale* 4-16, *P. malariae* 6-12). Merozoite buds appear around the schizont periphery, each receiving a nucleus that buds off from the central mass containing the DV full of HZ crystals. During this stage intense synthesis and assembly of molecules that are needed for RBC invasion also occurs. Merozoites eventually pinch off from the residual body of cytoplasm, the RBC membrane and the PVM lyse by a protease-dependent process and the merozoites exit into a brief extracellular phase before a new erythrocytic cycle begins with the invasion of fresh RBCs (8).

Sexual Forms

Concomitantly with the process of RBC invasion, a small portion of merozoites in newly invaded RBC differentiates into sexual forms, which are macrogametocyte (female) and microgametocyte (male); however, what triggers this alternative developmental pathway leading to gametocyte formation is unknown. When the *Anopheles* mosquito ingests the gametocytes, the sexual cycle of the parasite takes place (9).

Mosquito Stages

Once inside the mosquito, the gametocytes are taken into the midgut. Macrogametocytes escape from the RBC to form macrogametes, and microgametocytes exflagellate, each forming 8 haploid motile microgametes. The microgamete fertilizes a macrogamete and forms a zygote, which transforms into a motile ookinete. The ookinete penetrates the midgut wall and transforms into an oocyst. Ten to 24 days after infection, depending on the *Plasmodium* species and ambient temperature, thousands of motile sporozoites are released into the hemocoel and invade the salivary gland epithelium. When an infected mosquito bites a susceptible vertebrate host, the *Plasmodium* life cycle begins again (9).

MALARIA: THE PROBLEM

Distribution and Global Situation

Malaria is the most important parasitic disease worldwide, ranking among the major health and developmental challenges nowadays (18). There are 300 to 500 million clinical cases every year, and between 1 and 3 million deaths -mostly of children under the age of 5- are attributable to this disease (19). The global pattern of malarial transmission suggests a disease centered in the tropics, but with a reach into subtropical regions in 5 continents (18). Although malaria remains an important health problem in Asia and South America, its main impact is in sub-Saharan Africa, where at least 90% of deaths from malaria occur. The main reason for the persistence of malaria in Africa is the presence of the vector *Anopheles gambiae*, although social and economical factors are also important. The *A. gambiae* complex of species has a remarkable tendency towards human biting (antropophily) and is long-lived, making it particularly effective at transmitting malaria (20). Such inoculation rates, combined with the long duration of parasite survival in the host, rapidly saturate local human populations, resulting in prevalence and superinfection (18).

The last century witnessed many successful programs at country level to eliminate the parasite; however, the world is now facing a rapidly increasing disease burden. This has been attributed to several causes, including insecticide resistance, the weakening of public health systems in poor countries, population movements into malaria-endemic regions due to war, and changes in agricultural practices (18, 20). However, by far the most important factor is the development of *P. falciparum* and *P. vivax* resistance to unexpensive and effective drugs, such as chloroquine (CQ) and sulphadoxine/pyrimethamine (21). With a rapidly growing population in regions of high malaria transmission, it has been estimated that in the absence of effective intervention strategies the number of malaria cases will double over the next 20 years (19).

Control Strategies

Different approaches are currently used to control malaria in endemic areas: (i) Vector control through the use of insecticides, mainly DDT, which is still a valuable control tool in some situations when used in household spraying (22). (ii) Interruption of human-vector contact with insecticide-treated bednets (ITBs) that provide significant protection; unfortunately, resistance to insecticide has now been recorded in different regions of Africa (23). (iii) Improving access to treatment as a result of training the population (24). (iv) Combination therapy to improve drug efficacy and to decrease drug resistance by the parasite; however, the benefits of this treatment must be balanced against the draw-back of increased costs (25).

The goal of reducing morbidity and mortality caused by malaria requires the implementation of an integrated program. Major points that need special attention include reduction or elimination of contact between humans and infective mosquitoes, early diagnosis and identification of individuals at risk of developing severe malaria, rapid identification and monitoring of drug resistance, drug development and new therapeutics, and vaccine development. Of interest, the *Plasmodium* genome project is identifying potential new drug targets at an accelerating rate (26), and this knowledge may allow a drug already in use for some other purpose to be developed as an antimalarial quickly and relatively at low cost (21). Moreover, exploitation of the *Plasmodium* genome project will facilitate vaccine development, and the *Anopheles* genome project may help to identify novel insecticides, ways of overcoming insecticide resistance and facilitate attempts to develop mosquitoes that are resistant to malaria infection (26). Importantly, improvement in the healthcare systems and a major investment in staff recruitment, training and support after training are needed if the new tools that are being developed for malaria control are to be introduced and evaluated effectively (20).

Treatment

Prompt and effective treatment is probably the most cost-effective element of malaria control. The bulk of antimalarial therapy worldwide is oral drugs for uncomplicated *P. falciparum* malaria. This treatment prevents progression to severe disease, and, if the drugs are efficacious, they reduce overall malaria morbidity and mortality (27). However, most people living in endemic areas have little or no access to diagnosis and treatment. Adequate malaria treatment also requires that effective and safe drugs are protected against the emergence of resistance (25). Drugs with long half-lives and for which resistance is conferred by point mutations select resistant parasites rapidly (28).

Malaria had been sheltered for many years from the dangers of resistance because of the outstanding properties of CQ and the slow speed at which resistance developed to this drug (29). The final arrival of resistance to CQ on a global scale led to the development of other drugs such as the antifolates (30). Unfortunately, malaria control authorities now face the challenging situation of global resistance against both CQ and the antifolate sulphadoxine/pyrimetamine, the two most widely used antimalarials (21). Over-reliance on the same classes of drugs resulted in cross-resistance among closely related chemical entities (*e.g.,* quinolines and antifolates). Moreover, sequential introduction of drugs as monotherapy led to sequential selection and spread of mutant drug-resistant parasites, and ultimately to multidrug resistance (31).

Combinations of artemisinin-derivatives -a group of promising antimalarial endoperoxides - with quinolines or antifolates provide mutual protection against resistance, high efficacy, and reduce transmissibility. However, their cost is elevated and the safety of these combinations needs to be further assessed. In addition, further work in dosing, co-formulation and delivery is urgently needed (32). It is therefore of paramount importance to determine the best alternative therapies available for use now as well as the best prospects for drug development and their regulatory approval and use in short term (21). The targets of the main asexual blood-stage antimalarials are related to the functions of distinct parasitic organelles. Of particular interest are the pathways related with parasite metabolism that differ significantly from the human host, such as the process of heme detoxification. In addition, some of these drugs exert their antiparasitic effects on other phases of the parasite life cycle, including liver stages (quinolines and antifolates) and sexual-stage differentiation (artemisinin-derivatives, quinolines, antifolates, atovaquone) (31). With the sequencing of the *Plasmodium* genome and the technological advances in drug discovery, it is hoped that new classes of antimalarials will be developed (33).

Vaccines

With an increase in insecticide and antimalarial drug resistance, considerable effort has been undertaken during the last two decades to develop a malaria vaccine; however, its design is extremely challenging because of the complexity of *P. falciparum* life cycle and the variety and severity of disease syndromes that it causes (34). To appreciate the constraints of developing a malaria vaccine is important to consider that in contrast to most acute infectious diseases -for which sterile immunity is reached following natural infection- in malaria there are several immune responses that restrict parasite growth, but the parasite persists (26). Moreover, if immune responses lead to chronic infection, the parasite benefits from such responses, which thereby enhance transmission to the mosquito (34).

Further complicating the picture, the parasite expresses stage-specific antigens (8) and many parasite proteins exhibit polymorphism (35), which potentially limits the effectiveness of a malaria vaccine. In addition, repeated infections with heterologous *Plasmodium* species occur and according to previous studies, protection against different *Plasmodium* species appears to be difficult to achieve with a single vaccine (36). Thus, once successful vaccines against *P. falciparum* and *P. vivax* –the two species that contribute the most to malaria burden- have been developed, the possibility of combining antigens into a single formulation will have to be assessed (34).

Richie and Saul (34) reviewed some important points that should be taken into account in order to develop an efficient antimalarial vaccine. Importantly, selected antigens must be accessible to the immune system and induce protective immune responses. Moreover, antigens from different stages as well as several antigens from a single stage should be combined. However, it is important to keep vaccines as simple as possible to facilitate formulation, to optimize the immunogenicity of the various components and to reduce costs. Mutistage, multigene DNA-based vaccines against *P. falciparum* -currently in development - may offer a way to increase antigenic diversity (37). Finally, because biological characteristics of each parasite stage determine which immune responses are optimal (*e.g.,* antibodies [Abs], CD8- and/or CD4-expressing T-cell responses), the vaccine must induce the right kind of immune response (38, 39). Thus, particular adjuvant systems, such as CpG, may also be needed to drive responses in a given direction (40). Methods for screening specific-stage proteins that induce protective immune responses are already being developed (33). The sequence of the parasite genome and an increasing understanding of its biology, should greatly contribute to vaccine development because it should allow selection of the best candidate antigens (34).

Immune Responses to Malaria: Protection and Pathology

Naturally acquired immunity (NAI) to malaria takes as long as 10-15 years of exposure to develop and, even with repeated infection, protective immunity is non-sterile. In endemic areas, individuals frequently have premunition, (*e.g.,* parasitemia and Abs without symptoms) (41). Their immunity to infection is exposure-related and therefore age-related, and occurs rapidly to severe non-cerebral disease, more slowly to mild disease and probably never to asymptomatic infection (42). NAI to malaria involves both Ab-mediated and cell-mediated immunity, and is species-, strain-, variant- and stage-specific (43).

Immune Responses to Liver-stage Parasites

Immunization with irradiated sporozoites has been shown to elicit protective immunity not only in animal models, but also in human volunteers (44). Consistent with these studies, individuals living in endemic areas, who are naturally exposed to malaria, also develop immune responses to sporozoites. The anti-sporozoite Ab titers gradually increase throughout childhood and by early adulthood the majority of individuals are seropositive (45). Early studies, using monoclonal Ab (mAb) specific for CSP repeats, provided the first evidence that Ab alone could protect against sporozoite challenge (46). In addition to humoral immune responses, the first direct evidence indicating that T cells play an important protective role against pre-erythrocytic forms was obtained from experiments showing that sporozoite-immunized mice depleted of $CD8^+$ T cells exhibited a significantly reduction in their sterile immunity (47). Similarly, transfer of $CD4^+$ T cell clones derived from sporozoite-immunized mice demonstrated that these cells strongly inhibited parasite development and conferred protection (48); however, the underlying mechanisms responsible for this protection have not been completely defined. The complexity and variability of the immune responses to liver stages depending on the murine strain and parasite species limits our ability to extrapolate these findings to identify protective effector mechanisms in sporozoite-immunized humans. Nevertheless, the clear demonstration that sporozoite-induced humoral and cellular immunity can protect *in vivo* provides the rationale for multiple approaches to vaccine development (49).

Immune Responses to Asexual Blood-stage Parasites

Innate Immunity

Theoretically, following acute *P. falciparum* infection in humans, it would be expected that all RBC be infected within 10-12 days; however, in practice, parasitemia is only ~0.1% (50), suggesting that innate mechanisms control parasite growth before antigen-specific immunity develops. Monocytes/macrophages (Mϕ), neutrophils (Nϕ) and NK cells are the main players in this type of immunity (51, 52). Enhanced phagocytic activity by monocytes/Mϕ and rapid parasite clearance from circulation by the reticuloendothelial system is a feature in patients infected with *P. falciparum*, *P. vivax* and *P. ovale* (53). The antiparasitic effect of human monocytes/Mϕ via phagocytosis has been reported on *P. falciparum* merozoites and other asexual intraerythrocytic forms (54, 55). Whereas direct phagocytosis of free parasites or iRBC reduces the parasite load, blood monocytes or tissue Mϕ can also mediate parasite inhibition by their capacity to produce reactive oxygen intermediates (ROS) and reactive nitrogen intermediates (RNI) (56-59). In addition to monocyte/Mϕ activation by NK cell-derived cytokines, Mϕ are directly activated by iRBC and by parasite components, such as GPI and HZ. Such activation triggers diverse effector mechanisms, including cytokine production (60-65), NO (66) and ROS (67), which exert antiparasitic effects. Similarly, direct activation of dendritic cells (DC) (*e.g.*, cytokine and MHC II up-regulation) has been detected in response to *P. chabaudi*-iRBC (68) and to HZ (69). Nϕ-mediated phagocytosis of *P. falciparum* was observed in infected individuals (70), and Nϕ were shown to kill *P. falciparum* intraerythrocytic asexual forms as well as free merozoites *in vitro* (71). NK cells increase in numbers during infection and lyse *P. falciparum*-iRBC *in vitro* (72). Because of their ability to lyse target cells without prior antigen sensitization in a non-MHC-restricted fashion, NK cells are considered to be one of the cell types participating in innate immunity against malaria. In addition to their cytolytic activity, NK cells have been shown to produce IFN-γ and possibly tumor necrosis factor (TNF)-α, thereby leading to parasiticidal Mϕ activation in response to malaria infection (73).

The mechanisms through which the parasite induces this innate defense are poorly defined; however, different lines of evidence point to a role for toll-like receptor (TLR)-mediated signaling. Activation of Mϕ through TLR ends up in the induction of effector genes, whose products control and execute the innate defense in a large variety of bacterial and viral systems (74). Similarly, activation of TLRs (*e.g.,* TLR2, TLR4) in response to parasitic infections, including *Trypanosoma cruzi* (*T. cruzi)* (75), *Toxoplasma gondii* (*T. gondii*) (76) and *Leishmania major* (*L. major*) (77), appears to play an important role in host defense. Although this mechanism of immunity remains largely unexplored in malaria infection, two separate works have reported TLR-mediated signaling in response to *Plasmodium*. In mice infected with *P. berghei*, neither elevation of IL-12 levels nor liver injury was observed in mice deficient in myeloid differentation factor 88 (MyD88), an adaptor molecule shared by TLRs, indicating the requirement for TLR-MyD88 pathway in the induction of these inflammatory events (78). Moreover, a recent study demonstrated that *P. falciparum* activates DC via TLR9-MyD88-dependent signaling (79). These data suggest that the TLR system is also important in the innate defense against malaria infection. Further investigation will bring light on this important matter.

T cell- and Antibody-mediated Immunity

$CD4^+$ T cells are essential for immune protection against asexual blood-stages whereas for $CD8^+$ T cells this role is less clear (80). In experimental malaria, activation of parasite-specific $CD4^+$ T cells bearing αβ TCR can be initiated soon after they recognize malaria antigens in the context of MHC molecules on antigen-presenting cells (APC), such as Mϕ (81) and DC (82). It has been shown that $CD4^+$ T cells alone are able to confer protection against *P. chabaudi*. Severe combined immunodeficient (SCID) mice reconstituted with T cells from immune donors suppressed parasite growth, suggesting that these cells contribute to immunity. Moreover, when $CD4^+$ T cells were depleted using an anti-CD4 mAb, mice lost the ability to control parasite growth (83). Nevertheless, several studies indicate that protective immunity to blood-stage in malaria requires both $CD4^+$ T cells and Ab (84). $CD4^+$ T cells have two major effector roles: (i) they act as helper cells for B cells, enabling the production of high affinity specific immunoglobulin (Ig)G Ab and the development of B cell memory, and (ii) they are able to activate Mϕ and other effector cells to produce IFN-γ, NO and ROS through the release of inflammatory cytokines (*e.g.* IL-12) (85). Both of these roles appear to have an impact on the control of blood-stage parasites (80). After the peak of parasitemia, which seems to be controlled by the Th1-type T-cell mechanisms, a switch to a Th2-type T-cell cytokine response ensues (86), resulting in clearance of parasites in the blood. This sequential activation suggests that early acute parasitemias are controlled by Ab-independent pathways, whereas later recrudescent parasitemias are controlled by Ab-dependent, B cell-mediated mechanisms (87). Thus, it appears that both Th1 and Th2 cells contribute to protective immunity at different times of infection, and the balance between these two subsets determines the outcome of the disease (39).

It is well established that B cells and Abs play a crucial role in immunity to malaria. NAI to malaria in individuals living in endemic areas, which takes several years to develop, is largely dependent on the acquisition of a repertoire of specific protective Ab directed against highly polymorphic blood-stage parasite antigens (*e.g.,* PfEMP1, RESA, MSP1, MSP2) (88). Treatment of *P. falciparum*-infected patients with IgG extracted from immune African adults resulted in reduction of parasite load and clinical symptoms (89). Similarly, passive transfer of mAb against parasite antigens conferred protection against *P. yoelii* infection in naïve mice (90). Mice lacking B cells were unable to clear parasites and instead developed a chronic infection that was resolved only by adoptive transfer of B cells from immune donors (84). In addition, this type of studies have revealed the essential role of B cells in switching from a Th1 response to a Th2 response, which is required for the complete resolution of *P. chabaudi* infection (84, 86).

Dysregulation of Immune Responses

Exacerbation of Immune Responses

Protective antimalarial immunity reflects cellular activities such as phagocytosis, Ab production, cellular cytotoxicity and parasite inhibition exerted by lymphocytes, Nϕ and monocytes/Mϕ; however, some of these cellular activities may also be harmful for the host (80). A number of inflammatory mediators have been proposed to be responsible for the pathogenesis of severe malaria. These include ROS, NO and cytokines, with the possibility of a dual role for some, since they appear to have both beneficial and detrimental effects in the host (91).

Reactive Oxygen Species

Clark and colleagues (92) proposed that ROS may play a role in the pathogenesis of severe malaria by causing endothelial damage. In humans there is some evidence that ROS participate in malaria pathogenesis. Monocyte-derived ROS generation was increased in adults with *P. falciparum* and *P. vivax* malaria, especially in those with severe disease (58). In children, ROS were associated with anemia (93) and products of lipid peroxidation were found in cerebrospinal fluid from CM patients

(94). Thus, even though Mϕ-derived ROS seem to be important for parasite control they might also contribute to the pathogenesis of malaria (56).

Nitric Oxide

In addition to the putative role of NO overproduction in CM (discussed below in the section "Cerebral Malaria"), the participation of NO in malarial immunosuppression has also been proposed (112).

Cytokines and Chemokines

Severe disease and high mortality associated with *P. falciparum* infection were previously attributed solely to parasitic virulence factors (*e.g.,* multiplication rate, antigenic variation, iRBC sequestration); however, a growing amount of evidence indicates that the host immunological response also contributes to malaria pathophysiology (95). In fact, high levels of inflammatory cytokines TNF-α, IFN-γ, IL-6 and IL-1β are more frequently observed in patients suffering from severe malaria than those suffering from mild disease or asymptomatically infected (96-98). Similarly, serum concentrations of chemokines IL-8 and (macrophage inflammatory protein) (MIP)-1α (99, 100), and myeloid-related proteins (MRP) S100A8/A9 (101) are also augmented during *P. falciparum* infection.

In addition to the overall magnitude of cytokine activation, the imbalance between pro- and anti-inflammatory responses appears to be predictive of fatal outcome. Death from malaria in adults was associated with high ratios of plasma IL-6 to IL-10 (98), whereas severe anemia in children correlated with high ratios of TNF-α to IL-10 (102). Thus, the amounts and the time at which these inflammatory mediators are released may be important determinants of subsequent pathophysiological events. Whereas their production may be protective early in malaria infection, prolonged high levels might be detrimental (91). Moreover, since erythrocytic schizogeny occurs in the microvasculature of organs where iRBC are sequestered, the immune response would presumably be directed against the sites of maximal parasitization, and the intensity and nature of this response may therefore contribute to the development of severe disease. Evidence from autopsy and *in vivo* histological studies indicates that in *P. falciparum* malaria, the main sources of cytokines are likely to be activated tissue Mϕ in liver (Kupffer cells), spleen, lungs (alveolar Mϕ) and brain (microglial cells), as well as PBMC (95). The putative roles of cytokines in malaria paroxysms and in the various complications related to severe disease -mainly CM, anemia, and PM- will be addressed in the section "Malaria: The Disease". Understanding the cytokine interactions that lead to both control and pathology will be helpful in the design of future immune treatment against malaria to prevent millions of deaths (103).

Immunosuppression

It is clearly advantageous for the parasite to develop methods to inhibit potentially harmful immune and inflammatory responses (104). Its ability to actively modulate the host immune system, preventing the development of specific immune responses might explain the poor acquisition of malaria immunity in naturally exposed populations (105). Several studies have shown impaired immune responsiveness in *P. falciparum* malaria. In humans, immunosuppression appears late in the acute phase of malaria and can last a long time after the clearance of parasites from the circulation (106). The immunosuppression that accompanies malaria is of practical importance, with malaria-infected children having more severe gastro-intestinal infections (107) and reactivation of chronic and latent viral infections (108). Likewise, there is evidence of this phenomenon in murine malaria (109).

Defects in immune reactivity include reduced T-cell proliferation, alterations in T-cell functions, decreased cytokine induction and suppression of Ab production (110, 111). Even though no single mechanism can explain the suppressive effects of malaria infection, several studies have suggested that defects in Mϕ accessory cell function are involved (*e.g.,* antigen presentation, Fc receptor-mediated RBC clearance) (106). Moreover, suppression of the lymphoproliferative responses of splenocytes to mitogens during malaria infection was associated with an immunosuppressive effect of NO and prostaglandin (PG) production (112).

In human and murine malaria, Mϕ play a prominent defensive role (106, 113). At the beginning of infection the number of Mϕ is dramatically raised and their activation is reflected by an increase in Fc receptor expression, iRBC and free parasite phagocytosis, cytokine and ROS production, and cell-mediated cytotoxicity (114-116). However, as infection progresses these activities decline, indicating Mϕ dysfunction (117). Concerning the parasite components responsible for these effects, the most marked immunosuppression has been described for crude schizont extracts, soluble exoantigens (67, 118) and for the parasite metabolite, HZ (119-122), as discussed in detail in Chapter 2.

In addition to Mϕ, DCs are efficient APC and therefore they are expected to be a target for parasite-mediated inhibitory actions. Indeed, *P. falciparum*-iRBC binding to the surface of myeloid DCs profoundly altered their maturation and down-regulated their functions. These effects were mediated *via* CD36 and CD51 ligation and included the inhibition of MHC II, adhesion molecule and costimulatory molecule (CD83, CD86) up-regulation in response to LPS, as well as reduced IL-12 secretion and increased IL-10 production (123, 124). Moreover, the ability of DC to stimulate allogenic and antigen-specific primary and secondary T-cell responses was also dramatically reduced (123). In contrast, murine DC matured normally and stimulated T-cell responses following exposure to *P. chabaudi*-iRBC. In addition, they produced TNF-α, IFN-γ and IL-12 (68), suggesting that at the cellular level differences exist between modulation of human and murine DC by malaria parasites, which might be related to variations on the DC receptors involved (104).

Altogether, these studies point to a role for parasite products and iRBC binding to phagocytic cells in malaria immunosuppression. If a direct effect is established and the underlying mechanisms are unraveled, then inhibiting these pathways of cellular damage may allow amelioration of morbidity and mortality related to malaria (104).

MALARIA: THE DISEASE

P. falciparum and, to a much lesser extent, *P. vivax* are the main causes of disease and death from malaria (125). In endemic areas, malaria infection results in a range of outcomes from asymptomatic infection, through mild disease, severe disease and death. A complex interplay between the parasite and the host, the microenvironment affecting the risk of exposure, and the ability of the host to develop a protective immune response are likely to account for this variability in disease pattern (91).

Mild Malaria

Most episodes of infection with *P. falciparum* in malaria-endemic areas lead to mild clinical symptoms, such as fever, malaise, headache, and lethargy followed by spontaneous recovery, which may occur even without drug therapy. Following repeated exposure to malaria immune responses eventually develop and protect against serious disease but do not convey sterile immunity. Low-grade infections occur with few or no clinical symptoms and, in a proportion of cases, severe disease results (126).

Important clinical manifestations of severe malaria include CM and profound anemia, as discussed below; however, these life-threatening complications occur in only a small minority of infections (127). In contrast, fever is an invariable feature of malaria infection in non-immune individuals. Even though fever is debilitating for the host, it is also beneficial since parasite growth is inhibited at febrile temperatures, thereby contributing to reduce parasite burden (128). Malaria fever is closely linked to the asexual erythrocytic cycle of the parasite. Indeed, there is now evidence that this fever is mediated by endogenous pyrogens produced by monocytes and Mϕ in response to parasite components released after schizont rupture (60-64, 129). As far as 1912, it was suggested that HZ was a factor in the production of the malaria paroxism (129). This work was neglected, but a more recent study showed that HZ exhibits TNF-α-inducing properties and thermoregulatory capacities (61). Similarly, the ability of GPI -another parasite component released after schizogeny- to induce TNF-α production and pyrexia has been demonstrated (60). In addition, a number of different malaria antigens have also been reported to stimulate TNF-α production, including RESA (62), a soluble antigen complex known as Ag7 (63, 64), and MSP1 and MSP2 (60); however, for these two latter antigens, the TNF-α-inducing ability was lost when the fatty acid was removed, suggesting that the activity resided in their GPI anchor moiety.

Accumulating data indicate that TNF-α is a critical mediator of malarial fever. When human monocytes were cultured in presence of *P. falciparum*-iRBC, the rate of TNF-α production augmented sharply at the time of schizont rupture (130); in *P. vivax* infection, fever paroxysms were associated with a marked increase in circulating TNF-α levels (131); and an anti-TNF-α therapy inhibited fever in children with CM (132). In addition to TNF-α, circulating levels of other pyrogenic cytokines (IL-6 and MIP-1α) have also been detected in malaria patients (96, 99); therefore, it is conceivable that they also participate in the induction of fever. It is difficult to dissect the precise contribution of each of these cytokines to clinical symptomatology, since their regulation tends to be highly coordinated. Future trials on anti-cytokine therapy may help to clarify this issue (133).

Severe Malaria

It is increasingly recognized that severe malaria is a disorder that affects several tissues and organs, even when the most marked manifestations may seem to involve a single organ, such as the brain (125). Clinical manifestations of severe infection are variable in nature and severity, and it can be difficult to clearly define syndromes in many cases (126). Individuals may suffer systemic, single- or multi-organ involvement, including CM, anemia, acidosis, acute respiratory distress, coagulopathy, shock, hypoglycemia, renal dysfunction or failure, pulmonary edema, and severe hemoglobinuria (134).

Severe malaria is complex and involves several key pathogenic processes. Such events include rapid expansion of iRBC mass, iRBC sequestration in the vascular endothelium, microvascular obstruction leading to hypoxia, and local release of inflammatory mediators. This in turn may lead to downstream events at a cellular level that further exacerbate the situation (126). When these general processes occur on specific organs, such as the brain and the placenta, they lead to the development of specific pathologies associated to severe malaria.

Cerebral Malaria (CM)

CM syndrome appears to account for the majority of deaths in non-immune patients and in infected African children. Mortality from CM remains at about 20%, even in hospitals with good facilities (20). CM is defined as an unrousable coma not attributable to other causes in patients with *P. falciparum* malaria. In African children, coma is very often accompanied by fever, convulsions, seizures, metabolic acidosis and hypoglycemia. A significant proportion of survivors have permanent neurological complications including cognitive and speech disorders, motor abnormalities and cortical blindness (135, 136).

Pathogenesis of human CM

During *P. falciparum* infection, sequestration of iRBC occurs within the microvasculature of the major organs in the body, including the brain. Parasite sequestration in the brain causes mechanical obstruction of blood flow leading to cerebral edema, raised intracraneal pressure, hypoxia and petechial hemorrhages, which were once thought to be the main cause of coma associated with malaria (16). However, it is now widely recognized that the production of proinflammatory cytokines also plays an important role in the progression of the cerebral pathology. Higher TNF-α levels were recorded in children who died of CM in two African populations (137, 138). Moreover, a report on postmortem human tissues from patients suffering of CM demonstrated induction of TNF-α and IL-1β in the brain (139).

As to the mechanisms through which proinflammatory cytokines participate in the development of CM, it has been postulated that this occurs *via* two main mechanisms: (i) the induction of intracellular adhesion molecule (ICAM)-1 expression in the microvascular endothelium of the brain, thereby promoting parasite sequestration in cerebral vessels (140); and (ii) the activation of iNOS to generate NO production from vascular endothelium; NO might then cross the blood-brain barrier and disrupt neurotransmission, inducing a state of unconsciousness (141). Whereas nitrate levels -as a marker for NO production in the peripheral blood and urine- revealed a positive association between NO and severe malaria (142), negative associations have also been reported

(143). In a separate study, high levels of iNOS expression were detected in brain tissue from fatal cases of CM (144), underlining the importance of monitoring local iNOS and NO production during CM.

Even though the cellular sources for inflammatory mediators in CM remains to be confirmed, different studies point to an important role for infiltrating monocytes and proliferating microglial cells (145, 146). In fact, a prominent feature of human CM is the activation of the microglia. Analysis of cellular activation in the brain revealed that these cells express MRP8 (S100A8) and MRP14 (S100A9) (147). Moreover, in *P. falciparum*-infected children, serum levels of MRP8/14 were significantly related to parasite load and fever episodes (101). Because these proteins are strong chemoattractants and activators (148), through their induction, microglial cells might be contributing to the inflammatory process in the brain. Besides direct parasite-mediated monocyte/Mϕ cytokine induction, studies of brain tissue form fatal CM cases suggests that specific anti-malaria IgE-containing immune complexes bind to activate monocytes to cause local overproduction of TNF-α, particularly in the deep vasculature where iRBC sequestration takes place (149). Thus, both direct and indirect parasite effects on mononuclear phagocytes may contribute to the development of CM.

In addition to a role for inflammatory responses in CM development, it is likely that metabolic disturbances also contribute to this pathology. In fact, individuals with profound coma can regain consciousness very rapidly, and may show no neurological sequelae, suggesting that the pathology of CM has a metabolic component (150). Acidosis and hypoglycemia have been associated with the worst outcomes in individuals with severe malaria infections (134, 151). Acidosis is largely metabolic, attributable in part to increased lactate production by the host and by parasite biomass; however, most important by far is the combined effects that reduce oxygen delivery to tissues (151). Hypoglycemia may arise from reduced hepatic gluconeogenesis, increased glucose consumption by a large parasite biomass (134), and/or might be triggered by parasite components such as GPI (60). Based on these studies, human CM appears to involve vascular obstruction and inflammation leading to tissue damage and disordered neurotransmission, as well as metabolic disorders. However, the sequence of events and causality can only be determined by more direct approaches. Experimental models designed to establish the relationship between neurological dysfunction and immunological and metabolic parameters are required to unravel the processes that lead to coma and death in CM (152).

Experimental CM (ECM)

The availability of different mouse strains has enabled to gain insight into the molecular mechanisms involved in CM. The parasite of choice has been *P. berghei* ANKA because of its ability to sequester within the microvasculature (152). *P. berghei*-induced CM is limited to strains of mice that are genetically predetermined to make a strong proinflammatory immune response (*e.g.,* CBA, C57BL), confirming the association in humans between proinflammatory cytokines and CM. These mice develop neurological signs and symptoms typical of human CM (*e.g.,* coma, vascular plugging, petechial hemorrhages) and die 8-10 days after infection (153). However, these models do not precisely reproduce all the features of human CM, since mainly monocytes -with a few iRBC- are sequestered in the brain, in correlation with enhanced vascular permeability and microvascular damage (154). Sequestration of iRBC in the brain was recently reported in (BALB/c x C57BL/6)F1 mice infected with *P. berghei* ANKA (155), and this was also described in the *P. yoelii*-BALB/c and -Swiss models (156, 157). Thus, despite some differences between human CM and ECM, the various murine models offer a useful tool to better understand, and eventually control, one of the major complications related to *P. falciparum* malaria.

Pathogenesis of ECM

Cytokine overproduction, in particular of TNF-α and IL-1β, appears to be involved in the pathogenesis of human CM (139). In line with these data, the principal mediator of ECM was identified as TNF-α (158), which seems to be required for NO production, adhesion molecule up-regulation, monocyte recruitment and microvascular damage during ECM (159). In keeping with human CM, ICAM-1 up-regulation was associated with monocyte and iRBC sequestration in the brain (155). Although TNF-α appears to be crucial for CM pathology, its production is regulated by other proinflammatory cytokines, including IFN-γ; thus, when IFN-γ responses were blocked (*e.g.,* anti-IFN-γ Abs, defects in IFN-γ receptor), mice were protected against CM (137).

Even though there are very few data available on the incidence of human CM in T cell-suppressed individuals, several lines of evidence point to a major role for T cells in the immunopathology of ECM (158, 160-164). T cell-deficient BALB/c mice were resistant to CM induced by *P. berghei* infection (160); both anti-CD4 and anti CD8 Ab treatments prevented ECM (158, 161); and brain-infiltration of αβ $CD8^+$ T cells correlated with the neurological symptoms of ECM and was responsible for mortality caused by *P. berghei* infection (162, 163). Further supporting this notion, CC chemokine receptor (CCR)-5 expression on both brain-sequestered leukocytes and nonhematopoietic cells was required for ECM development. An effect that was attributed to increased brain-sequestration of cytotoxic $CD8^+$ T cells (164). Taken together, these data indicate a pathological role for T cells in the development of ECM.

In addition to T cells, other leukocyte populations also appear to contribute to the pathophysiology of ECM. Indeed, Nϕ depletion in circulation prevented early development of ECM and dramatically decreased microhemorrhage and monocyte sequestration in the brain. Moreover, abrogation of IL-12p40 mRNA expression and down-regulation of TNF-α, IFN-γ and IL-2 transcripts in the brain was also detected (165, 166). In correlation with these data, peritoneal Nϕ and Mϕ from mice suffering from CM expressed various cytokine (*e.g.,* IL-18, IFN-γ, TNF-α) and chemokine (*e.g.,* MIP-1α, IFN-γ-inducible protein [IP-10]) transcripts (167). Altogether, these studies indicate that Nϕ and Mϕ may play a role in the pathogenesis of ECM via proinflammatory cytokine and chemokine expression. Other cells that have been implicated in cytokine-mediated cerebral pathology include microglial cells and astrocytes, which seem to contribute to TNF-α and chemokine (*e.g.,* MCP-1, IP-10) induction in the brain (168, 169).

<u>Severe Anemia</u>

Severe malarial anemia is probably the second most common complication of *P. falciparum* infection, but it can also be caused by *P. vivax* (125). Despite being a major cause of death in regions of high endemicity (170), the pathogenesis of malarial anemia is complex and mutlifactorial, and remains poorly understood. Severe anemia can be observed at low levels of parasitemia, during chronic infection, and even after complete parasite clearance (171). Several mechanisms have been implicated in the development of human malarial anemia: iRBC destruction at schizont rupture, accelerated removal of both iRBC and non-infected RBC, increased sequestration of iRBC, decreased RBC deformability (134), as well as ineffective erythropoiesis, bone marrow dyserythropoiesis, and lower erythroblast proliferative rates and numbers (172). The suppression of erythropoiesis in cases of severe malaria occurs despite an adequate production by the host of functional erythropoietin, the growth factor necessary for erythrocyte progenitor development (173). Similar observations have been made in murine malaria models (174, 175); however, the mechanistic basis for the suppression of erythropoiesis in the presence of erythropoietin is not completely understood.

Clark and colleagues (176) proposed that pathogenic manifestations of malaria, including severe anemia, may be due to proinflammatory cytokine secretion by host Mϕ in response to malaria parasites and their products. In different murine models of malaria, a soluble mediator released from the bone marrow and spleen cells was shown to depress *in vitro* erythropoietin-induced proliferation of erythroid precursors and to be partly responsible for anemia (177). Stevenson and colleagues (178) have shown that IL-12 seems to exert a protective effect against malarial anemia, but have ruled out the involvement of TNF-α, IL-1β and IFN-γ as erythropoietic inhibitors (179). In contrast, Migration inhibitory factor (MIF), detected in *P. chabaudi*-infected mice, inhibited erythropoiesis *in vitro* and its serum levels correlated with parasitemia and disease severity *in vivo* (180). These data suggest that MIF could be an important host factor in the development of severe anemia. Evaluation of effects of blocking MIF during murine malaria will help to validate this hypothesis.

Disease in Pregnancy

Among malaria-exposed adults, pregnant women, mainly primigravidae, are particularly susceptible to malaria. In areas of low transmission, malaria in pregnancy may lead to the death of the mother due to severe anemia, abortion of the fetus or a stillbirth (181, 182). In areas of high malaria transmission, heavy parasitization of the placenta may occur, a condition defined as placental malaria (PM) (183). A particularly striking observation during PM is the accumulation of vast numbers of iRBC in the placental blood spaces at much higher concentrations than seen in the peripheral blood. Substantial findings now point to the importance of specific receptor-mediated adhesion in the sequestration of iRBC in the placenta (184). Parasite sequestration in the placenta has been shown to impair placental function and results in intrauterine growth retardation, premature delivery and low birth weight (182). Because infant survival is so sensitive to even small reductions in birth weight, PM almost certainly increases infant mortality, although this has not been demonstrated directly (20).

In addition to the accumulation of iRBC in the intervillous spaces of the placenta, inflammatory cells such as monocytes/Mϕ frequently accumulate within the placental environment during PM (185). In line with these observations, PM infection was associated with elevation of mRNA levels of various chemoattractants and activators, including MIP-1α, MCP-1, I-309 and IL-8, which correlated with monocyte density in the placental intervillous space (186). Even though the presence of mononuclear phagocytes is important to control parasitemia, the inflammatory reaction leads to placental lesions associated with poor birth outcomes (187). Both protective and deleterious effects have been reported to be caused by the release of inflammatory mediators in the placenta. Elevated production of IFN-γ from intervillous blood mononuclear cells (IVBMC) was associated with protection against PM (188). However, overproduction of TNF-α and IFN-γ correlated with negative pregnancy outcomes, such as low birth weight infants (189). Similarly, increased expression of TNF-α and IL-8 by placental Mϕ was associated with intrauterine growth retardation during PM (190). The importance of cytokines in mediating the effects of malaria on pregnancy is consistent with the observation that *P. vivax* –which induces higher increases in TNF-α than *P. falciparum* (131)- also adversely affects birth outcomes (191).

Altogether, these studies indicate that infiltrated monocytes/Mϕ might contribute to the pathology of PM via the release of proinflammatory mediators and, probably, by further enhancing leukocyte recruitment into the placenta. In addition, it is possible that the presence of Mϕ in the placenta may favor HIV infection and vertical transmission, not only because they are potential reservoirs for Mϕ-tropic virus, but also because during PM, maternal Mϕ were found to express increased mRNA levels of CCR5 (192), one of the main coreceptors implicated in host cell invasion by HIV (193).

Parasite Components as Pathogenicity Factors

Because the clinical manifestations of malaria are associated with schizogeny and inflammatory mediator production (95), the potential contribution of GPI and HZ (two parasite products released during this process) has been investigated. Different lines of evidence indicate that both GPI and HZ have the ability to modulate immune cell functions (*e.g.,* NO, ROS and cytokine production) (60, 61, 65-67, 69, 119-122, 180, 194-205). As these immunological responses are implicated in malaria pathophysiology, these molecules appear as attractive targets for antimalarial therapy. A better understanding of their mechanism of action and their antigenic properties might eventually lead to the development of an antidisease vaccine aimed to reduce morbidity and mortality related to severe malaria. In this section of Chapter 1, studies regarding malarial GPI will be discussed, and Chapter 2 will focus on the characteristics and immunomodulatory properties of HZ.

Glycophosphatidylinositol Anchors

Structure and Functions

GPIs are a special class of glycolipids ubiquitous in eukaryotic cells. The primary function of GPIs is to anchor certain proteins to outer surfaces of cell membranes. In animal cells, GPIs appear to occur in low abundance and are predominantly found in the protein-anchored form. In parasites (*e.g.*, *Plasmodium, Leishmania, T. cruzi, T. brucei*), GPIs occur at relatively high levels, both as moieties that anchor proteins and as free GPIs (206). GPIs anchor functionally diverse proteins, which comprise molecules that provide barriers against immune attack, receptors and cell adhesion molecules. (*e.g., Leishmania* gp63, CD14). (207). In *P. falciparum,* GPIs serve as membrane anchors for several of the most important surface antigens of parasite invasive stages, including

MSP1 and MSP2 (208). The GPI moieties of a number of proteins have been implicated in important biological functions, including transmembrane signaling, apical targeting of proteins in polarized cells, insulin-mimetic activity, cell-cell interactions, embryogenesis, and development (207). The important functions of GPIs might explain the highly conserved basic structure of these molecules during evolution. However, parasite GPI structures are significantly different from those of humans, and therefore, their biosynthetic pathway might offer a specific target for the development of novel antimalarial drugs (206). Studies with inhibitors of this pathway have shown that this function is essential for *Plasmodium* development and survival (209).

GPIs represent the major carbohydrate modification in proteins of the intraerythrocytic stage of *P. falciparum* (210). The chemical structure and biosynthetic pathway of the malarial GPI anchor has now been elucidated, and comprises the structure NH-CH_2-CH_2-PO_4-(Manα1-2)6Manα1-2Manα1-6Manα1-4GlcN$H_2\alpha$1-6(acyl)*myo*-inositol-1-PO_4-diacylglycerol (211). This moiety is completely conserved among more than 20 geographically distinct isolates of *P. falciparum* examined to date (212). Free and protein-bound GPIs have also been identified and characterized in the rodent *Plasmodium* species, *P. chabaudi chabaudi* AS, which had almost identical patterns to those of *P. falciparum* (213). Similarly, the biosynthesis and characterization of GPIs in blood-stage forms of *P. yoelii yoelii* was recently reported (214). Even though the pathway leading to the formation of GPI precursors resembled that described for *P. falciparum*, in *P. yoelii* an additional modification was detected in the GPI structure. Subsequent *in vivo* studies of the effects of rodent parasite GPIs may provide a deeper insight into the relationship between malarial GPIs and the host immune response.

Activation of Immune Responses

It is clear that GPIs retaining the fatty acid moiety intact can leave the membrane and re-associate with other cells (215). Furthermore, GPI-anchored proteins can also be found in solution (216). Thus, it is thought that intact GPI may be released upon schizont rupture to interact with host tissues. Accumulating evidence indicates that plasmodial GPIs (60, 66, 195) as well those of other protozoans (217, 218) display biological activity, which might contribute to parasite-associated pathogenesis. Therefore, understanding of the GPI structure-activity relationship and cell signaling mechanism may offer insight into the development of new therapeutic strategies for malaria.

The carbohydrate nature of cytokine-inducing plasmodial antigens was first suggested by the demonstration that a heat-stable, protease-resistant, glycosidase-sensitive component in the crude extracts of *P. falciparum* led to TNF-α production in monocytes and Mϕ, and was able to kill mice hypersensitized to TNF-α (219, 220). Subsequently, Schofield and Hackett (60) demonstrated that either GPI-linked MSP1 and MSP2 or highly purified *P. falciparum* GPI induced high levels of TNF-α and IL-1 from peritoneal Mϕ; an effect that was blocked by a PKC inhibitor. Moreover GPI caused pyrexia and cachexia *in vivo*, and was shown to be insulin-mimetic, regulating glucose metabolism by adipocytes and inducing profound hypoglycemia *in vivo*. Finally, mice immunization with highly purified malarial GPI resulted in a serological anti-GPI response. The anti-GPI sera were able to neutralize TNF-α induction and lipogenesis by both the MSP1 antigen and whole parasites, suggesting that GPI plays an important role in TNF-α overproduction and hypoglycemia associated with malaria.

In addition to cytokine production, *P. falciparum* GPI was found to up-regulate adhesion molecule expression (ICAM-1, VCAM-1 and E-Selectin) in human vascular endothelial cells (HUVEC). Moreover, plasmodial GPI increased *P. falciparum* and peripheral blood mononuclear cells (PBMC) cytoadherence to HUVEC. Both GPI-mediated effects were blocked in presence of protein tyrosine kinase (PTK) and nuclear factor κB (NF-κB) inhibitors, indicating that endothelial cell activation by GPI is PTK- and NF-κB-dependent. Parasite-induced ICAM-1 expression by HUVEC was abrogated by an anti-*P. falciparum* GPI mAb (195). These data suggested that the release of GPI in the local environment could contribute to iRBC and leukocyte cytoadherence by activating the vascular endothelium, and thereby favoring microvascular obstruction. Further supporting a role for plasmodial GPI in malaria immunopathology, iNOS expression and NO production were detected in murine peritoneal Mϕ and HUVEC stimulated with *P. falciparum* GPI. In Mϕ, this effect was linked to PTK, protein kinase C (PKC) and NF-κB activation since inhibitors against these second messengers reduced GPI-mediated NO release. In addition, an anti-GPI Ab neutralized NO induction by *P. falciparum* extracts *in vitro*, suggesting that GPI is an important parasitic agent in NO induction, and therefore might play an important role in CM (66).

These observations concerning Mϕ activation by plasmodial GPI are in agreement with other studies reporting the proinflammatory properties of GPIs from different parasitic origins (217, 218, 221, 222). PTK-mediated IL-1α and TNF-α induction by GPI derived from the membrane-bound variant surface glycoprotein (mfVSG) of *T. brucei* was reported in murine Mϕ (221, 222). GPIs isolated from *T. cruzi* triggered NO synthesis and microbicidal activity in IFN-γ primed Mϕ (217). It was recently demonstrated that the potent TNF-α-, IL-12- and NO-inducing activities of *T. cruzi* trypomastigote mucins resided in their GPI anchor fraction (218). These reports concur that GPIs constitute important inflammatory molecules in these diverse parasite systems. In contrast to *P.*

falciparum and *Trypanosoma* GPIs, the main *Leishmania* GPIs –lipophosphoglycan (LPG) and glycoinositolphospholipids (GIPLs)- were shown to inhibit Mϕ functions, such as NO production (223) and PKC-mediated *c-fos* expression (224). These data demonstrate the specificity of action among GPIs from various protozoan taxa, suggesting that these molecules may serve differentially to regulate host cell function during protozoan infections (225).

As to the mechanisms underlying these regulatory events, it appears that parasite GPIs can function both as agonists and as substrates in cell signaling through two distinct structural elements, the glycan and the lipid moieties, respectively. The conserved glycan core was reported to activate PTK while the diacylglycerol moiety was a second messenger for PKC activation. Thus, it has been proposed that the immune responses induced by *P. falciparum* GPI require dual action of the carbohydrate and the lipid moieties in the covalently linked form (225). Subsequent structure-activity analyses on *T. cruzi* and *P. falciparum* GPIs supported this conclusion (218, 226). Even though very little is known about how plasmodial GPIs are able to cause outside-in activation of cell signaling pathways, it is now clear that GPIs are neither inserted to the Mϕ plasma membrane nor internalized. *P. falciparum* GPI is rather recognized by the Mϕ *via* its distal fourth α-mannose residue (226). This interaction represents a novel mechanism in cell signaling involving initial recognition of this residue by a mannose-specific protein, presumably localized in the Mϕ membrane in association with PTK.

Altogether, the studies indicate that parasite GPI anchors can activate signaling pathways in host cells and lead to the expression of inflammatory mediators, thereby modulating the immune response. In the case of malaria, this cellular activation might contribute to the immunopathology related to the disease, including cytoadherence, immune dysregulation, hypoglycemia and CM.

Plasmodial GPI as Target for Immunotherapy

In malaria endemic areas, adults and children older than 4-5 years have protective immunity, acquired in response to repeated infection with *P. falciparum* over a period of time (26). Naik and colleagues (227) reported that this resistance was associated with the development of circulating Abs directed against the parasite GPI anchor; thus, providing the first evidence for immunological recognition of this parasite component by humans. Whereas adults and older children had high levels of persistent anti-GPI IgGs, younger children either lacked or had low levels of short-lived anti-GPI Abs. The absence of a persistent anti-GPI Ab response correlated with malaria specific illness (*e.g.,* fever, anemia), while the presence of long-lived anti-GPI Abs in sera paralleled the naturally acquired resistance to malaria pathogenesis, suggesting that anti-GPI Abs could provide protection against clinical malaria (227). Although the Abs were mainly directed against the acylated phosphoinositol portion of GPIs, much needs to be done to validate GPI as a target antigen; this is, to determine the provenance and epitopic specificity of these Ab responses. Among other considerations, there is insufficient data at present as to non-self discrimination by anti-GPI Abs to determine whether they could be administered immunoprophylactically. Along with the potential for GPI as antidisease vaccine, these areas are currently under investigation (228).

Further supporting the notion that GPIs may be useful for active or passive immunization against malaria, Schofield and colleagues (229) recently reported that immunization of mice with a chemically synthetic version of the core GPI glycan of *P. falciparum* provided significant clinical protection against the lethal *P. berghei* ANKA CM syndrome. Recipients were substantially protected against malarial acidosis, pulmonary edema, cerebral syndrome and fatality. In addition, anti-GPI Abs neutralized proinflammatory activity by *P. falciparum in vitro*. These data indicated that GPI is causally involved in rodent malaria pathogenesis, including metabolic derangement, and that vaccination against this target confers clinical protection in this model. Thus, this study provides an initial model for the further development of an antidisease vaccination against malaria. However, whether GPI is a target of clinical immunity in humans remains to be determined (229). Detailed knowledge of the structure-activity and structure-immunogenicity relationship of GPI may offer insight into the development of GPI-based therapeutics or a vaccine for malaria disease (206).

CHAPTER 2
HEMOZOIN

HISTORICAL BACKGROUND

Hemozoin (HZ) or malaria pigment was previously named "plasmodin" and also "hemo-melanin" until Sinton and Ghosh (230) credited Sambon with the origin of the term "hemozoin" in 1934. This pigment was described in malaria patients long before the parasitic nature of this disease was clarified. In 1717, Lancisi was the first to note discoloration of the internal organs in malaria victims (231). In 1847, Meckel identified the characteristic pigment in the viscera and blood of people who died of "pernicious fever" and proposed that the pigment itself was the cause of malaria (232). Though Meckel was in error, HZ served as a visible marker that contributed to the identification of the malaria parasite and the elucidation of its life cycle. In 1849, Virchow was the first to connect malaria infection with the presence of pigment in the blood (233) and, in 1850 Hischl associated it with intermittent fevers (234). However, it was only thirty years later, when Laveran discovered the malaria parasite itself, that the source of HZ became clear (235). In his description of unstained smears from malaria patients, he noted that pigment granules were visible inside parasites. Confirming and extending these observations, Councilman and Abbot noted the blackish color of brain, lungs, liver and spleen of patients who died of malaria and associated it with the presence of HZ both inside and outside iRBC (236). Importantly, in 1897, when Ross detected the pigment inside bodies (parasites) contained in insects fed on malaria-infected blood, HZ helped to the identification of the *Anopheles* mosquito as the vector of malaria (237). Early this century, Brown identified the chromogenic species as hematin (238) and suggested that HZ might be involved in the pathophysiology of malaria (129, 239). Since then, a long and continuing debate regarding HZ structure, synthesis and function has persisted.

HEMOZOIN STRUCTURE

Despite its early identification, the chemical and structural nature of HZ has only been elucidated in the last decade. As early as 1911, Brown distinguished the malaria pigment from melanin (238) and hypothesized that during the digestion of Hb by *Plasmodium* there was enzymatic discoordination of ferrous heme (Fe^{II}-protoporphyrin [PP])IX) from globin, and thus HZ was equivalent to hematin (HO-Fe^{III}-PPIX). During the 1930s and 1940s numerous investigations with a wide variety of *Plasmodium* species supported the view that HZ was crystalline hematin. Common features between HZ and hematin included their chemical composition (carbon, nitrogen, iron and oxygen content), as well as their identical spectra when HZ was dissolved in strong basic solutions (240, 241). However, by dissolving HZ in mildly basic solvents, Deegan and Maegraith demonstrated that the hematin structure was incompatible with the solubility properties of HZ -which is highly insoluble compared to hematin- and suggested instead that HZ comprised heme in association with an unidentified protein (242). It was then proposed that this protein, which co-purified with HZ, could either be denatured globin (243) or a parasite-derived polypeptide (244).

Debate continued on the composition of HZ until 1987 when Fitch and Kanjananggulpan demonstrated that pigment purified by protease digestion still was intact HZ (245). These data confirmed that HZ contained only heme, which was sufficient to maintain its structure. In 1991, Slater *et al.* (246) first characterized HZ and synthetic HZ (sHZ), β-hematin, using infrared and X-ray spectroscopy. They established that the infrared spectra of both the native pigment and the synthetic preparation were essentially identical, and identified the unique iron to oxygen bond that distinguishes HZ from non-crystalline hematin and its aggregated states. This bond was found between the central ferric ion of one oxidized or ferric heme (Fe^{III}-PPIX) and the carboxylate side-group oxygen of another one. Their structural prediction for both HZ and β-hematin was a step-ladder polymer $(Fe^{III}\text{-PPIX})_x$ (Figure 1A) (246, 247) although no direct evidence for this last proposal existed. In 1997, Bohle and co-workers (248) first reported a high-resolution powder diffraction pattern of both native HZ and β-hematin. The high quality diffraction data were collected using synchrotron radiation and permitted determination of the unit cell dimensions of the crystal. Both materials corresponded to a single crystalline triclinic lattice in which each unit cell is occupied by two Fe^{III}-PPIX molecules that are symmetry-related via a center of inversion. These results unambiguously demonstrated that HZ present in *Plasmodium* is β-hematin. Subsequently, similar data were used to determine the chemical structure of β-hematin (249). This study provided definitive evidence that β-hematin is a crystalline cyclic dimer of Fe^{III}-PPIX, which involves reciprocal coordination of the ionized heme-propionate side chain of one Fe^{III}-PPIX to the Fe^{III} center of the other. These dimeric units interact in the crystal *via* intermolecular forces that include hydrogen bonding through the remaining un-ionized heme propionic acid group on each Fe^{III}-PPIX (Figure 1B) (250). Thus, the language and concepts of polymerization are not appropriate to understanding HZ formation (251) because this material is not a polymer as previously believed, but is rather indicative of a crystal biomineralization process, as it will be discussed below. Even though both native HZ and the synthetic HZ are formally β-hematins (248), for clarity here, the *in vitro* synthetic preparations are called β-hematin or sHZ and the native pigment is called HZ or malaria pigment.

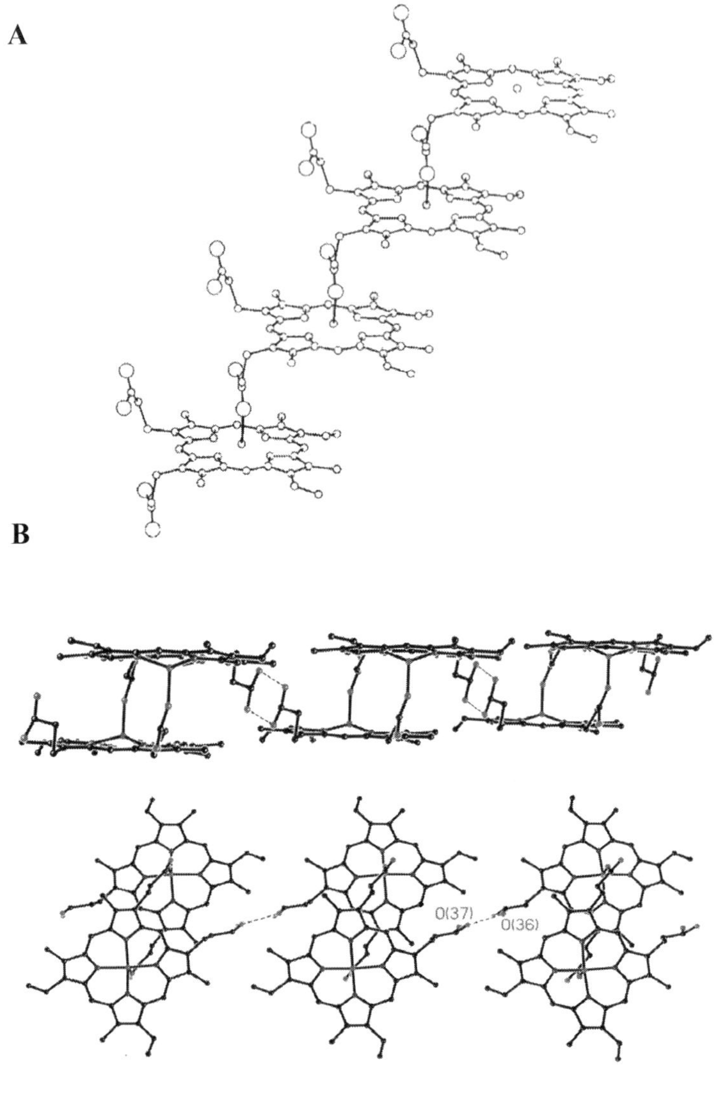

Figure 1. Chemical Structure of Hemozoin. (A) Model of a step-ladder HZ polymer. Reprinted from *Experimental Parasitology*, Vol 74(3):362-365, Slater, A. F. , Malaria pigment, Copyright 1992, Academic Press Inc, with permission from Elsevier. Reference (247). (B) Two views of the dimeric structure of β-hematin showing the iron-oxygen and propionic acid linkages between dimeric pairs of hemes. Reprinted from *Acta Crystallographica Section D Biological Crystallography*, Vol D58:1752-1756, Bohle, D. S., A. D. Kosar, and P. W. Stephens, Phase homogeneity and crystal morphology of the malaria pigment beta-hematin, Copyright 2002, with permission from The International Union of Crystallography (IUCr). http://journals.iucr.org/ Reference (250).

Hemozoin Synthesis

During the intraerythrocytic phase of its life cycle, the parasite *Plasmodium* matures within a cell in which Hb is the single major cytosolic protein (95% of the RBC cytosol). The malaria parasite has a limited capacity to synthesize amino acids *de novo* (252) or to take them up exogenously (253). Therefore, while in the trophozoite stage, it avidly ingests and degrades between 60% and 80% of host erythrocyte Hb into amino acids for use in protein synthesis and as an energy source (252). The parasite ingests Hb by means of a specialized mouth-like structure called cytostome, which sucks RBC cytosol into the parasite by a process resembling pinocytosis (254). Hb-containing vesicles are pinched off from the cytostome and carry their contents to the DV, an acidic organelle (pH 5.0-5.5) where Hb is broken down (255). This process releases heme, which does not appear to be metabolized or recycled but rather accumulates within the DV as crystalline particles called HZ (256). In contrast to the trophozoite, during the ring stage, Hb digestion appears to be limited; however, HZ can still be observed, suggesting that the cellular machinery for ingestion and proteolysis is present (257). In fact, several studies have revealed that early in parasite development small portions of cytoplasm are taken up by micropinocytosis and, as the parasite matures, a larger volume of Hb is ingested via the cytostomal system (254).

Hemoglobin degradation

To account for the enormous amount of protein breakdown, which occurs in just a few hours of the trophozoite stage (258), it was postulated that the parasite must possess an efficient and probably highly specific pathway for Hb proteolysis. Several attempts to study this process were hampered by difficulties in purifying plasmodial proteases and by the inability to define which proteases were truly involved in Hb digestion. Various putative hemoglobinase activities were described in different species of *Plasmodium* (259, 260), and both cysteine (261) and aspartic (262) proteases were proposed to be the major hemoglobinolytic enzymes. However, it had not been possible to demonstrate that any of these activities had a role in Hb proteolysis *in vivo*.

In 1990, Goldberg and colleagues (263) showed that Hb degradation in the malaria parasite *P. falciparum* occurs in the DV, a specialized acidic organelle that has evolved to catabolize Hb, to utilize its amino acids and to coordinate heme into HZ in an efficient manner. Isolation of parasitic vacuoles allowed the identification of enzymes present in this organelle, which are capable of Hb proteolysis. Among the multiple and diverse proteases at work in this organelle, aspartic proteases accounted for 60-80%, and cysteine protease(s) for 20-40% of the globin-degrading activity (263, 264). Proteolysis in the vacuole was found to be an ordered process, requiring an aspartic protease activity to clip intact Hb in large fragments before other proteolytic activities can function efficiently.

Two vacuolar aspartic proteases have been purified and characterized. Aspartic hemoglobinase I or plasmepsin I is the one that initiates Hb degradation by cleaving it in the hinge region that maintains its quaternary structure, thereby unraveling the molecule for further proteolysis. The second one, hemoglobinase II or plasmepsin II, is an aspartic protease capable of cleaving Hb with an overlapping specificity, but seems to prefer denatured Hb (264). Even though it is not known why these two different enzymes are so similar in their specificity, it is possible that they act in concert to increase the efficiency of Hb breakdown. Plasmepsin I is synthesized during both the ring and trophozoite stages, whereas plasmepsin II is almost undetectable in rings but is abundant in trophozoites. These observations suggest that only plasmepsin I is involved in the limited Hb proteolysis during the ring stage, whereas both proteases may be required to expedite this massive catabolic effect during the trophozoite stage (265).

A third vacuolar protein, falcipain, was identified as a cysteine protease, which cannot degrade native Hb without reducing conditions but readily cleaves denatured or fragmented Hb (264). From these studies it was concluded that falcipain likely works by degrading Hb fragments after initial aspartic protease attack. In contradiction to the above results, in another study, which utilized reducing conditions, falcipain was shown to cleave native Hb (266). Moreover, in studies with cultured parasites, cysteine but not aspartic protease inhibitors caused a block in Hb degradation (261), suggesting that plasmodial cysteine protease activity contributes to initial steps in Hb degradation. However, as discussed by Goldberg (267), it is possible that the fairly non-selective cysteine protease inhibitors studied acted in other important cellular process that could disrupt the vacuolar conditions necessary for Hb degradation. Cell fractionation and immunolocalization experiments will be critical to define the role of this protease. In conclusion, though the precise order of action of the three identified food vacuole proteases (plasmepsins I and II, falcipain) remains to be confirmed *in vivo*, it appears that these enzymes -and probably other peptidases- act in concert to digest Hb to free amino acids.

Exactly when and where Hb degradation starts is also a subject of controversy (267). On the one hand, it has been suggested that newly formed vesicles from the cytostome begin to breakdown Hb, digest their second membrane and eventually coalesce into one residual DV (268). Alternatively, it has been proposed that Hb digestion starts after the vesicle is delivered into the DV since undigested Hb is seen enclosed in one membrane inside the DV (269). Moreover, fusion of preexisting DV with the outer membrane of Hb-filled vesicles has been observed, followed by lysis of the vesicle and Hb hydrolysis (270). Even though these observations support Hb digestion in the DV, they do not exclude the early degradation model; indeed, both processes may occur (267). Interestingly, in a later study, Hempelmann and colleagues (271) have put forward a challenging hypothesis on the digestion of Hb. Based on data generated by electron microscopy (EM), the authors suggested that this process occurs in endocytic vesicles named transport vesicles (TV)- before delivery to the parasite DV. However, several physiological constraints have been identified with this hypothesis, including the increase in the osmotic pressure inside the TV generated by the large concentrations of proteolytic products, and the rapid membrane lysis by the predicted high local concentrations of free heme (272). Thus, more definitive proof is required before Hb digestion in the TV can be considered, at least, as an alternative mechanism to the efficient processes observed in the DV. As suggested by Goldberg twelve years ago (267), immunolocalization studies of the proteases involved in Hb degradation and the proton pumps implicated in acidification will be helpful in sorting out when and in what form a vesicle loaded with Hb is capable of initiating the catabolic process.

Crystal Formation

The process of HZ formation is of great interest to malariologists, not least because it is thought to be the target of some of the most effective antimalarial drugs (273). Even though the exact point in Hb digestion at which this process occurs has not been established (267), it is known that during this degradative pathway heme is released and immediately autoxidized to potentially toxic ferric heme. Free heme can damage cellular metabolism by inhibition of enzymes (274), peroxidation of membranes (275), and the production of oxidative free radicals in the acidic environment of the DV (276); therefore, its efficient disposal is of crucial importance to parasite survival. In *Plasmodium sp.*, rather than excreting or degrading heme, as it occurs in mammalian cells, the parasite has evolved a novel pathway for its detoxification by incorporating it into an insoluble, crystalline dark-brown substance called HZ that is harmless to the parasite (277). Morphologically, the crystals exhibit a triclinic habit and are uniform in shape and size (Figure 2) (271). These characteristics are typical of a process called "biologically controlled crystallization" -biocrystallization- in which an organism exerts a great deal of control over the nucleation and growth of the crystal (278). Even though it is widely recognized that a nucleation center or template is required to initiate the process, direct observation of HZ synthesis within the parasite is extremely challenging; thus, there is currently no consensus on the mechanisms of HZ formation *in vivo* (279).

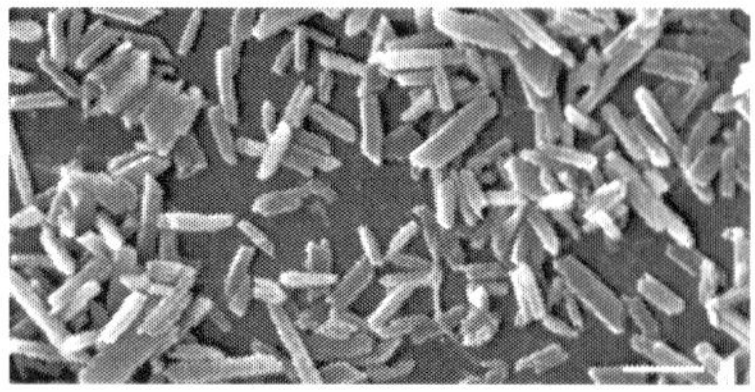

Figure 2. *P. falciparum* Hemozoin Crystals. Scanning electron micrograph of HZ crystals isolated from mature trophozoites of *P. falciparum*. The crystals appear as long needles (dimensions approximately: 1.0 μm X 0.6 μm X 0.2 μm). Scale bar = 1 μm. Reprinted from *Trends in Parasitology*, Vol 19(1):23-26, Hempelmann, E., C. Motta, R. Hughes, S. A. Ward, and P. G. Bray, *Plasmodium falciparum*: sacrificing membrane to grow crystals?, Copyright 2003, with permission from Elsevier. Reference (271).

Early studies indicated that HZ did not form spontaneously from either free heme or Hb under physiological conditions (245, 246). Consistent with this, in 1992 Slater and Cerami (15) reported that HZ synthesis was promoted by malaria parasite components and proposed the requirement for a specific enzyme "heme polymerase". Incubation of an extract of isolated *P. falciparum* trophozoites with ^{14}C-heme as the substrate resulted in conversion of the radiolabeled heme into HZ. The activity was associated with a membrane-pellet fraction, suggesting the involvement of a DV membrane-associated enzyme. HZ formation was dependent on the amount of membrane protein added, and was also time-, temperature- and pH-dependent, consistent with enzymatic activity. The pH optimum for the activity was found between 5.0 and 6.0, which covers the estimated pH of plasmodial DV (280). In a separate investigation, Chou and Fitch (281) also identified a "heme polymerase" activity in extracts of another species of *Plasmodium*, *P. berghei*, further supporting the hypothesis of Slater and Cerami.

In direct contradiction to the above results, it was demonstrated that HZ could be formed chemically -named β-hematin- in a test tube at 60-70 °C in an excess molar concentration of acetate and milimolar concentrations of monomeric heme (282). Extending these observations, Dorn and colleagues (283) described heme crystallization as an autocatalytic chemical process rather than an enzymatic one, which was dependent on the presence of heme-derived material associated with HZ and not on protein. This active principle was extractable with acetonitrile and resistant to destruction by proteolysis or heat, raising doubts as to whether HZ formation was enzyme-mediated. Moreover, either β-hematin or protein-free HZ subfractions from DV preparations incubated at 37 °C were shown to promote HZ synthesis. These observations indicated that preformed HZ could nucleate the production of further HZ. However, the authors hypothesized that an initiator, probably a protein, was required to begin the formation of HZ crystals in the parasite, working as a structural focus and, once crystals were present, further HZ growth could occur onto this nucleus (283).

Conversely, other groups were unable to detect HZ synthesis promoted by preformed β-hematin (284) and two independent investigations (285, 286) reported that under physiological conditions HZ formation was not spontaneous. Additionally, in the latter study (286), a heat-labile (resistant to proteolysis) and a heat-stimulable (inactivated by proteolysis) "heme polymerase" activity were identified in extracts of *P. berghei.* These data were interpreted as a plausible explanation for previous observations indicating HZ formation even when parasite extracts were submitted to high temperatures or to proteolysis (283). Thus, debate began over a spontaneous chemical reaction or a parasite directed process in the DV. Nonetheless, it is generally accepted that some component present in the parasite seems to be responsible for HZ synthesis, as the spontaneous reactions are too slow to account for the presence of such large quantities of HZ *in vivo* (287). As reviewed by Egan (279), in addition to the above-described hypothesis of HZ formation autocatalyzed by preformed HZ (283), two other hypotheses currently enjoy support. The first one is catalysis or initiation by histidine-rich proteins (HRP)s (288), and the second one is HZ formation catalyzed or initiated by lipids (289).

The first hypothesis is based on the initial study of Sullivan and colleagues (288) who demonstrated that heme crystal formation was supported by either native or recombinant HRP II. *P. falciparum* HRP II was shown to be present in the DV, and to bind heme at acidic pH and initiate HZ formation *in vitro*. Similarly, a related protein *P. falciparum* HRP III was found to bind heme and initiate HZ formation. Independently, other groups also demonstrated heme crystal initiation with *P. falciparum* HRP II (290, 291). Even though field isolates lacking *P. falciparum* HRP II have not been described, a laboratory strain that possess HRP IV but lacks both HRP II and III was still able to synthesize HZ. Given that *P. falciparum* appears to have redundant HRPs, at least two of which can mediate heme crystallization, it was concluded HRP II is sufficient but not absolutely necessary in this process (288, 292).

Supporting the hypothesis that HRP II promotes HZ formation, biomimetic peptide dendrimers corresponding to a repeat sequence (Ala-His-His-Ala-His-His-Ala-Ala-Asp) present in this protein were shown to bind heme and support HZ production in a time- and concentration-dependent manner (293). In a separate study Pandey and colleagues (294) found that repetitive hexapeptides of this sequence bound heme and inhibited HZ synthesis promoted by *P. yoelii* extracts, possibly by competitively binding free heme. These data suggested that such repetitive sequence represents the heme-binding site of plasmodial HRP II and probably the site of nucleation for HZ formation. However, for initiating heme crystallization some additional sites of HRP II may be necessary, as the repeat peptides alone could not induce HZ formation. Whether HRP II is actually involved in HZ synthesis in the parasite is not certain. Whereas confocal microscopy has demonstrated that HRP II is present in the DV, far larger quantities appear to be present in the RBC cytosol (291). Thus, it is not clear why a protein, which supposedly has the specific function of producing HZ would not be targeted directly to its site of action (279). In addition, the fact that preformed HZ can promote further HZ synthesis (283) suggests that HRPs may be more important in the initiation of HZ formation than in their extension. Detailed kinetic analysis will be required to establish whether HRPs function by binding substrate and releasing product, or whether they play a nucleation role (288).

The second hypothesis postulates that conditions in the DV are such that lipids independently of proteins can initiate HZ synthesis. This is based on the work of Bendrat and colleagues (289) who reported that phospholipids (*e.g.,* methyl esters of oleic, palmitic and stearic acids) contaminating both *P. falciparum* HZ and β-hematin preparations could cause rapid heme crystallization in the absence of protein and may well form a micelle that can orient the heme groups appropriately *in vitro*. According to these data, it is conceivable that phospholipids may account for the acetonitrile-soluble, heat-stable and protease-resistant "heme polymerase" activity previously described by Dorn and colleagues (283); however, it is unclear whether heme aggregation within *Plasmodium*-iRBC is effected solely by endogenous lipids. Since HZ forms much faster in the parasite than *in vitro*, it is thought that the appropriate pH in the DV and an adequate environment, possibly involving lipids, can account for this efficient process.

In line with the study of Bendrat *et al.* (289), membrane phospholipids were shown to be the most efficient "seeds" to start the biocrystallization of ferric heme (295), and several purified, unsaturated lipids from cell-free preparations of *P. berghei*-iRBC promoted HZ formation (296). This effect was attributed to the ability of lipids to solubilize heme. Thus, it was postulated that lipids co-precipitate with heme in the parasite DV and dissolve sufficient monomeric heme to allow HZ formation. As proposed by Hempelmann and co-workers (271), given that heme is virtually insoluble in acidic aqueous milieu but is highly soluble in lipid membranes, accelerated crystal growth could be promoted by virtue of the hydrophobic environment and by providing a template to which specific proteins can act as nucleation sites. In agreement with this reasoning, in a separate study, Pandey and colleagues (297) reported that HZ formation *in vitro* occurred more efficiently in presence of both *P. falciparum* HRP-II and lipids than when they were added separately, suggesting that HZ synthesis is a coordinated two-step process involving both lipids and scaffolding proteins. In this context, heme crystallization would be achieved by the action of HRP II in dimer formation, and linkage of heme molecules together with lipids. Lipids would in turn facilitate hydrogen bonding of two heme molecules and stabilize the structure by shielding these hydrogen bonds from competitive hydrogen binding with water.

Until now, the procedures for the assay of HZ formation do not represent the physiological conditions in the parasite, since *Plasmodium* degrades almost 80% of the host erythrocytic Hb and detoxifies the resulting heme in a few hours. Our knowledge of HZ synthesis is associated with *in vitro* procedures that provide 1-3% efficiency of heme crystallization, which is not sufficient to explain the actual amounts of HZ produced by the parasite to avoid toxicity. The data collected so far suggest that autocatalysis as well as HRPs and lipids are implicated, but their exact role (if any) *in vivo* remains elusive (278). Involvement of lipids and proteins is typical of many biomineralization processes, as is the growth of material onto existing crystal surfaces (279). Indeed, there are some interesting parallels between HZ formation and mechanisms of biocrystallization in other microorganisms. For example, in magnetotactic bacteria, the crystalline magnetic particles are surrounded by membranes, which are thought to be involved in the formation of the crystalline phase while also providing physical constraints on the size and shape of the crystal (298). Thus it is possible that all three processes may occur together (279). Analysis of factors responsible for HZ formation and reconstitution of these factors to mimic this physiological activity in the parasite is important to gain insight about this pathway, which is the target of major drug development efforts.

Hemozoin Formation by Diverse Plasmodium Species and other Organisms

HZ is present in all *Plasmodium* species with the exception of a single laboratory clone of *P. berghei* cultured under constant CQ pressure, which apparently transports either heme or HZ out of the iRBC (299). In a compartive study Noland and colleagues (292) analyzed the three-dimensional morphology and chemical properties of HZ from human (*P. falciparum, P. vivax, P. ovale, P. malariae*), primate (*P. knowlesi*, *P. brasilianum*), murine (*P. yoelii*) and avian (*P. gallinaceum*) *Plasmodium* species (Figure 3). HZ of the human *Plasmodium* species *P. vivax, P. ovale* and *P. malariae* were very similar to that of *P. falciparum*. Moreover, the crystals were quite stable in different chemical environments. In contrast, different morphologic shapes were observed between HZ from mammalian and avian *Plasmodium* species. Whereas HZ from human, primate and murine *Plasmodium* exhibited a regular, flat-faced cuboidal morphology with modest size differences, the avian pigment presented a larger, irregular barrel shape with a waffle surface. These distinct morphological characteristics could reflect potentially different biochemical environments, such as Hb and salt concentrations, and vacuolar pH, as well as distinct mechanisms of formation. In addition, invasion of a nucleated RBC in the case of *P. gallinaceum* is a host difference that may possibly contribute to the different appearance of the pigment. Even though morphological variations were detected in this study, it should be noted that no evidence for any difference in chemical structure was presented. X-ray diffraction data would be required to prove that structural differences exist among the various HZ obtained from different sources (279).

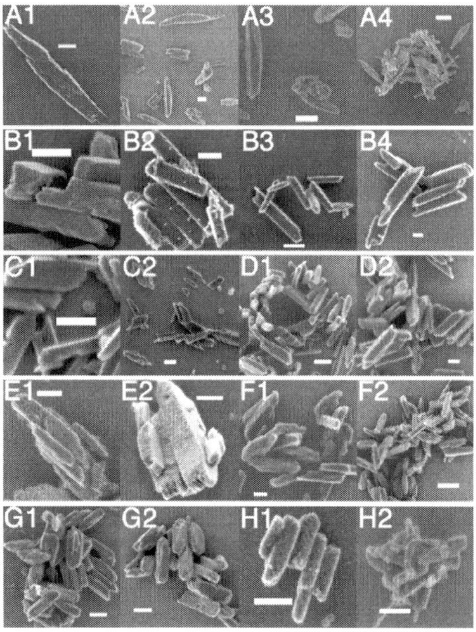

Figure 3. Hemozoin Crystals from Different *Plasmodium* Species. Electron micrographs of (A) β-hematin, (B) *P. falciparum* HZ, (C) *P. vivax* HZ, (D) *P. ovale* HZ, (E) *P. malariae* HZ, (F) *P. brasilianum* HZ, (G) *P. knowlesi* HZ, and (H) *P. yoelii* HZ. The size bar is 200 nm. Reprinted from *Molecular and Biochemical Parasitology*, Vol 130:91-99, Noland, G. S., N. Briones, and D. J. Sullivan, Jr., The shape and size of hemozoin crystals distinguishes diverse *Plasmodium* species, Copyright 2003, with permission from Elsevier. Reference (292).

In addition to *Plasmodium*, HZ formation has also been detected in other organisms. In 1999, Olveira and colleagues (300) were the first to demonstrate that HZ formation was not exclusive to *Plasmodium* but was also produced in the midgut of a blood-sucking insect, *Rhodnius prolixus* (*R. prolixus*). This hemiptera is an important vector of *T. cruzi* (the causative agent of Chagas disease), which ingests several times its own body weight of vertebrate blood and digests it within a few days after the meal (301). Thus, it is likely that HZ formation constitutes a strategy of the insect to detoxify heme produced during Hb degradation. In addition to HZ synthesis, *R. prolixus* possesses a heme-binding antioxidant protein (302), suggesting that HZ formation might be the first line of defense against heme toxicity, but not the only one developed by this organism. In *R. prolixus*, heme aggregation activity was promoted by an intestinal membrane fraction and was shown to be heat-labile (303). These results indicated that in this blood-sucking insect, there is a factor capable of promoting HZ formation, which may be analogous to the putative "heme polymerase" activity described in *P. falciparum* (15). However, the component responsible for stimulating HZ production in *R. prolixus* remains elusive.

The parasitic worm *Schistosoma mansoni* (*S. mansoni*) (helminth), the main etiological agent of schistosomiasis, digests large amounts of Hb in the host circulation to complete its development and sexual maturation (304). During this process, heme derived from Hb breakdown is released inside the worm gut. It was postulated that this parasite continuously regurgitated heme as a dark-brown material called *Schistosoma* pigment (SP) (305). Comparison of crude purified *Plasmodium* HZ and SP by EM led to the conclusion that the structures of these pigments were distinct (306). In a later study Oliveira and colleagues (307) characterized SP and unequivocally identified the isolated pigment as HZ. Similar to *Plasmodium* and *R. prolixus*, *S. mansoni* produces HZ in order to detoxify heme from Hb degradation. Indeed, HZ synthesis in *S. mansoni* seems to occur early in development (307).

Hemoproteus species, close relatives of *Plasmodium*, are apicomplexan protozoan parasites common in birds, and are relatively nonpathogenic. In contrast to *Plasmodium*, *Hemoproteus columbae* (*H. columbae*) is transmitted by flies rather than mosquitos, undergoes asexual schizogeny in endothelial cells of the lung rather than RBC, and only gametocyte invades RBC, which are nucleated (308). Chen and colleagues (309) reported that this parasite is able to digest Hb and detoxify heme by forming HZ crystals. In the same study, comparison of *H. columbae* pigment with HZ of *P. falciparum*, *S. mansoni* or synthetic HZ revealed that all four pigments served as templates for HZ extension and had similar infrared spectra, indicating that they have common unit linkage. However, they could be distinguished by their three-dimensional morphologies. Both *P. falciparum* and *H. columbae* HZ had symmetrical shape unlike the amorphous *S. mansoni* pigment and β-hematin (Figure 4) (310). These differences are suggested to be due to effects of intracellular versus extracellular HZ formation; however, the effect of initiating proteins or lipids in the final three-dimensional crystals is yet to be determined.

The finding of HZ synthesis into these widely different organisms suggests that other hematophagous animals may solve the problem of large amounts of potentially toxic heme by following the same strategy. However, according to the currently available data, many parasites digest Hb as an amino acid source, but only a few produce HZ instead of catabolizing heme *via* heme oxygenase (HO). In this regard, it is interesting to note that in contrast to *Plasmodium* and *Hemoproteus*, *Babesia* -an intraerythrocytic parasite related to *Plasmodium*- degrades little if any Hb and does not form HZ (311), as it is also the case for *Entameba histolytica* (*E. histolytica*), a protozoan that phagocytes RBC (265). Similarly, preliminary results from the Oliveira group (303) indicated that other hematophagous organisms such as mosquitos (*Anopheles* and *Aedes*) and the hard tick (*Boophilus microplus*) do not produce HZ. In contrast, others have reported the presence of pigments derived from Hb digestion in two Platyhelminthes, *Diclidophora merlangi* and *Fasciola hepatica* (312). It thus appears that the capacity to synthesize HZ may not be a universal solution employed by hematophagous organisms to detoxify heme, but rather might be associated with the chemical environment of the digestive process (303). In fact, *Plasmodium*, *Hemoproteus*, *R. prolixus* and *S. mansoni* digest Hb in an acid environment, and it is noteworthy that the process of heme aggregation is strongly pH-dependent (247). Thus, HZ formation appears to be an important feature in the adaptation of some organisms to hematophagy, which deserves further investigation in order to find specific targets for chemotherapy against pathogenic parasites.

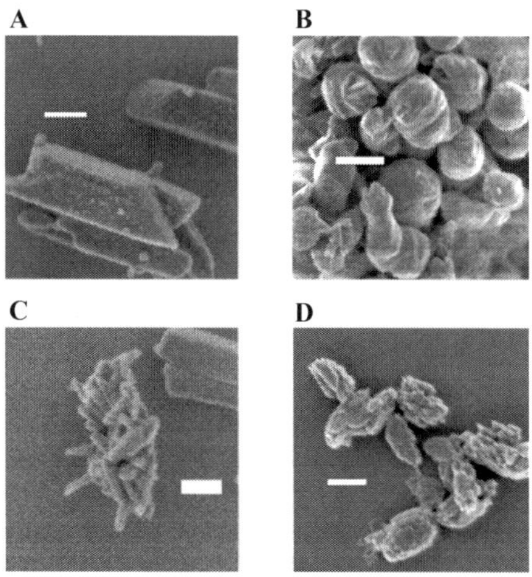

Figure 4. Hemozoin Crystals from Different Organisms. Electron micrographs of (A) *P. falciparum* HZ. Scale bar is 100 nm. (B) *S. mansoni* HZ. Scale bar is 300 nm. (C) β-hematin. Scale bar is 300 nm. (D) *H. columbae* HZ. Scale bar is 200 nm. Reprinted from *International Journal of Parasitology*, Vol 32:1645-1653, Sullivan, D. J., Theories on malarial pigment formation and quinoline action, Copyright 2002, with permission from Elsevier. Reference (310).

Synthetic Hemozoin (β-hematin) Formation

Isolation of HZ from *Plasmodium* is readily accomplished; however, it requires substantial quantities of cultured parasites and extraction of such amounts is impractical (279). Therefore, in addition to the studies that have attempted to unravel the process of heme crystallization inside the parasite, others have focused on the synthesis of HZ under entirely non-biological conditions; stated another way, on the formation of β-hematin or synthetic HZ (sHZ). Importantly, β-hematin has been shown to be chemically (245, 246), spectroscopically (246, 313) and crystallographically (248) identical to native HZ. Thus, investigation using large quantities of β-hematin has greatly facilitated the study of parasitic HZ.

Several methods have been reported for the preparation of β-hematin, which can be divided in two broad types, aqueous and non-aqueous. The aqueous method initially described by Slater *et al.* (246) involves heating HO-Fe^{III}-PPIX in 4.5 M acetic acid at 70 °C overnight, followed by washes in water, SDS and carbonate buffer (pH 10). Later Egan *et al.* (282) developed a modified version of this preparation in which hemin (Cl-Fe^{III}PPIX) is dissolved in 0.1 M NaOH to produce a solution of hematin. After neutralization of the NaOH with HCl during which the hematin precipitates, acetate is added to give a final concentration of 4.5 M acetate (pH 4.5) and this preparation is stirred and heated at 60 °C for 30 min to 1 h, followed by product washes in water. These data indicate that acetate facilitates β-hematin synthesis by increasing the solubility of hematin, thus behaving as a phase-transfer catalyst, a role that *in vivo* could be fulfilled by carboxylic acids or amino acids. In fact, amino acids (0.4 M glycine together with 0.1 M glutamate) could replace acetate in the reaction mixture.

Even though these observations led to the proposal that HZ formation in the parasite might be spontaneous, requiring no catalyst or initiator, subsequent investigations indicated that the synthesis of β-hematin in low acetate concentrations was very slow and seemed to rule out spontaneous HZ formation in presence of low concentrations of biological carboxylates (285, 295, 314). Indeed, one of these studies suggested that β-hematin was not formed at all in acetate solution at pH 4.5, proposing instead that the product was a heme-acetate complex (285). This was later refuted on the basis of X-ray diffraction, infrared spectroscopy and elemental analysis, which definitively showed that the product was β-hematin (315). Subsequently, it was suggested that β-hematin formed under these conditions was not produced during incubation of hematin in acetate solution, but rather during the drying process used to prepare the sample for infrared spectroscopy (316). This was also conclusively refuted on the basis of the infrared spectrum and X-ray diffraction pattern of the wet product, which was characteristic of β-hematin (251). Thus, β-hematin is certainly formed in acetate solution at moderately low pH and provides a useful model for probing the mechanism of HZ formation (279).

In parallel, Bohle and Helms (317) devised a strictly non-aqueous method, by which β-hematin is prepared from hemin by dehydrohalogenation with the organic base 2, 6-lutidine in rigorously dry methanol. In a later study Bohle *et al.* (250) examined the Egan and the Bohle β-hematin preparations by scanning EM and X-ray diffraction. Whereas both products were spectroscopically identical, major differences were found in the diffraction and morphological features of the precipitates. Even though the Egan crystals were similar to HZ from mammalian *Plasmodium* species, they were smaller and less regular in shape. In contrast, the Bohle crystals -with rectangular dimensions, smooth sides and far fewer tapered ends- most closely resembled to those formed by mammalian *Plasmodium* species. These results underline the importance of scrutinizing heme aggregates precipitated by different systems, especially by powder diffraction, to ensure that they are β-hematin and not some other insoluble phase (250).

In a separate study, Egan *et al.* (251) presented a detailed quantitative investigation of the kinetics of β-hematin formation in acetate solution, in which the effects of temperature, acetate concentration, pH, stirring rate and seeding were analyzed. The formation of β-hematin was found to occur *via* rapid precipitation of amorphous (or possibly nanocrystalline) hematin, followed by slow conversion to crystalline β-hematin. The reaction exhibited a behavior that was interpreted as evidence of a process of nucleation and growth phases, which would occur *via* dissolution of the initial hematin and re-precipitation as β-hematin. This type of behavior closely parallels many mineralization processes that take place by rapid precipitation of the most soluble solid (often amorphous), followed by slow conversion to increasingly less soluble, but thermodynamically more stable solids. By this reasoning the formation of β-hematin was paralleled to a biomineralization process.

On the basis of this evidence, Egan and colleagues (279) have proposed two possible models to account for HZ formation *in vivo*. If the release of heme from Hb is so fast that the concentration of hematin exceeds its solubility limit, it will precipitate. The process is then likely to proceed in a manner similar to that in acetate solution. There are indications that some carboxylates (Egan *et al.*, unpublished data cited in (279)) and lipids (295) could play a role analogous to acetate. On the other hand, if the release of heme from Hb is carefully controlled such that the solubility limits of hematin are never exceeded the formation of HZ could occur directly from solution and might require a suitable nucleation site. As previously described, both *P. falciparum* HRP II (288) and pre-formed HZ (283) are good candidates to act as a nucleation site for rapid growth of further HZ. In fact, the notion that HZ formation occurs under careful biological control and by interaction with a protein matrix is in accord with what is commonly observed in biomineralization processes. Experiments aimed at detecting whether large quantities of precipitated hematin are present in the parasite should be able to distinguish between these two models (279). In summary, the studies on β-hematin formation should be considered complementary to those evaluating the formation of HZ in presence of biological material. Although some of the presently available reports are contradictory, such information is valuable in the attempt to understand the mechanism of HZ formation *in vivo*, which still remains elusive.

Antimalarial Drugs Related to Hemozoin Synthesis

A variety of chemical compounds affect the processes of Hb digestion (*e.g.*, aspartic and cysteine protease inhibitors) and HZ formation (*e.g.*, quinolines, xantones) in the malaria parasite. Others appear to interact with free heme but their antiparasitic effects do not seem to be related to the inhibition of HZ synthesis (*e.g.*, endoperoxides). Understanding the mechanisms of action of these drugs is of paramount importance in order to identify new specific targets and to design synthetic or semisynthetic compounds for antimalarial chemotherapy.

Antimalarial Drugs that inhibit Hemoglobin Degradation

A number of inhibitors of the two classes of plasmodial vacuolar proteases -aspartic and cysteine proteases- have been shown to exhibit antimalarial effects. The aspartic protease inhibitor pepstatin affected primarily rings and schizonts and caused the formation of pyknotic parasites (318). The cysteine protease inhibitor E-64 had a major effect on trophozoites leading to the accumulation of large quantities of undegraded globin in DV and to eventual halt in parasite development (261). Of interest, the antimalarial effects of E-64 and pepstatin were markedly synergistic, suggesting that a combination therapy might be a promising approach (319). The plasmepsin inhibitor Ro 40-4388 has been identified as a potent inhibitor of Hb digestion, which effectively reduced the intracellular pool of heme (320). A specific peptidomimetic inhibitor, SC-50083, impaired Hb degradation by plasmepsin I (a plasmodial aspartic protease) *in vitro*, and when added to cultures at the trophozoite stage, it markedly reduced HZ formation and killed cultured parasites at low micromolar concentrations (321). Peptide library synthesis directed at the inhibition of plasmepsin II has allowed the identification of selective blocking compounds against this aspartic protease (322).

In a search for non-peptide protease inhibitors, a database of commercially available molecules was screened against the active site of falcipain, a plasmodial cysteine protease. This approach identified a lead compound as a potential inhibitor, and subsequent chemical synthesis and screening resulted in a set of analogues effective against cultured parasites at micromolar concentrations (323). Vinyl sulfones were shown to be potent irreversible inhibitors of falcipain. When the most potent of them was tested against cultured parasites, it blocked Hb degradation and impaired parasite development at nanomolar concentrations, suggesting that it could be the basis for future drug development (324). Moreover, peptidyl vinyl sulfones, which are regarded as less toxic than other cysteine inhibitors (*e.g.*, peptidyl fluoromethyl ketones), were more effective in delaying disease progression and cured 40% of *P. vinkei*-infected animals (325).

Altogether, these studies indicate that the approach of targeting parasite proteases is a promising avenue and may lead to the development of new antimalarial drugs. As recently discussed by Ridley (21), malaria chemotherapy might benefit from the development of inhibitors against a variety of proteases including, human immunodeficiency virus type 1 (HIV-1) aspartic protease, the cysteine proteinase cathepsin K (implicated in osteoporosis) and the aspartic protease rennin (an enzyme associated with hypertension). The challenge to chemists is to identify simple, non-peptidic structures with good oral bioavailability and low costs.

Antimalarial Drugs that Interact with Heme

Quinolines

Quinoline antimalarials owe their origins to quinine, an active ingredient of *Cinchona* bark, which was first imported into Europe from Peru for antimalarial use in the 17th century. Quinine has liabilities associated with toxicity and, because it requires a three-time daily administration over seven days, this can result in poor compliance (21). The dependence on raw material for its extraction and the opportunities presented by its structural elucidation led to the development of the fully synthetic 4-aminoquinoline antimalarials, notably CQ and later amodiaquine (AQ) (326). AQ use has been limited since the mid 1980s after it was linked to occasional agranulocytosis in travelers taking the drug prophylactically; however, because AQ retains a high degree of efficacy against the most highly CQ-resistant strains, there has been a recent increase in its use (327). Mefloquine and halofantrine are structurally related drugs that are active against CQ-resistant strains, but resistance can develop rapidly against each of them (29). Additional limitations include occasional neuropsychiatric disturbances for mefloquine, and halofantrine is strongly contra-indicated in people with a history of heart disease (21).

CQ was first introduced after World War II. This drug remains a major first-line antimalarial drug in many parts of the world, despite the loss of its effectiveness due to parasite resistance. CQ success was based on its safety, affordability, easy use and (prior to resistance development) to its efficacy (328). There is much literature describing the aspects of CQ pharmacology that are considered to be crucial to drug action (329); however, until now the exact antiparasitic mechanism of CQ is not fully understood. The related quinolines, quinine and mefloquine, which have been less studied than CQ, may act by some of the same mechanisms proposed for CQ (330).

Among the numerous modes of action ascribed to CQ –which will be discussed below- it has been postulated that this drug exerts its antiparasitic effects by inhibiting HZ formation (15, 282, 283, 288, 331-334). Some of the main findings that have led to this proposal include: (a) CQ inhibition is specific for parasites that are actively degrading Hb (331); (b) CQ and its congeners accumulate in the *P. falciparum* DV, where they hyperconcentrate to near micromolar concentrations from nanomolar concentrations in the plasma. As a neutral weak base, CQ is able to cross membranes but when protonated in an acidic environment, it becomes membrane impermeable leading to its accumulation (331); (c) EM autoradiography and subcellular fractionation has demonstrated that CQ binds to HZ in the DV of *P. falciparum* (332); (d) *in vitro*, HZ formation is blocked under approximately physiological conditions by CQ regardless of whether extension is mediated by crude trophozoite extracts, seed crystals of HZ, or β-hematin (15, 282, 283); (e) initiation of HZ synthesis by purified *P. falciparum* HRPs is inhibited by CQ (288); (f) quinolines bind measurably to HZ only in presence of heme. They do not bind directly to the HRPs, nor do they interact with isolated HZ in absence of crystal elongation. (332); (g) CQ binds heme non-covalently through π-π interactions of their planar aromatic structures (333); (h) binding assays demonstrated that the association quinoline-heme complex with the HZ crystal is specific, saturable, and of high affinity, and that diverse quinoline analogs can compete for binding (334).

From these studies, it is clear that the extension phase of HZ crystal formation is interrupted by the quinolines, whereas initiation may or may not be inhibited. As reviewed by Sullivan (310), two hypotheses have been proposed to explain the mechanism of inhibition of HZ synthesis by CQ. According to the first one, CQ binding to heme sequesters heme substrate from incorporation into HZ. Both free heme and the formed CQ-heme complex are toxic to the parasite (335); as a result, Hb digestion is impaired and parasite maturation arrested. The second one is that the CQ-heme complex incorporates into the growing crystal. This caps further extension, leading to the interruption of HZ formation and to the accumulation of toxic free heme, which in turn initiates the irreversible demise of the parasite (332). Recently, kinetic studies showed that inhibition of heme crystal growth by CQ and quinidine is reversible, suggesting a reversible association of drug-heme complex on growth sites of the crystal (336).

Independently of whether the CQ-heme complex blocks or not further crystal formation, both proposals predict an increase of heme delivered to the DV membrane causing damage by lipid peroxidation and increased membrane permeability, eventually leading to cell lysis (335). Heme can also bind enzymes, thereby inhibiting their action (260). Similarly, according to both theories, the effects of CQ would end up with more heme delivered to the parasite cytosol where reduced glutathione (GSH) degrades heme to release iron (310). In addition to heme binding, CQ and AQ were shown to competitively inhibit GSH-dependent heme degradation, allowing heme to accumulate in membranes (337). In agreement with the predictions of this model, CQ resistance in *P. berghei* and *P. falciparum* has been associated with increased cellular levels of GSH whereas sensitization of the parasites to CQ can be reached by reduction in GSH levels (338). Moreover, the *in vivo* antimalarial action of CQ and AQ was potentiated by drugs that reduce cellular GSH levels (*e.g.* indomethacin, acetaminophen) (339). Thus, CQ (and probably also AQ) appears to exert both direct and indirect effects on heme detoxification that are detrimental to the parasite.

In addition to preventing the formation and growth of HZ crystals, other mechanisms have been proposed to explain the antiparasitic effects of CQ. These include: (a) intercalation into parasitic DNA (340), (b) alkalinization of DV after accumulation due to its weak base effects (280), (c) inhibition of heme-dependent protein synthesis (341), (d) prevention of iron release from Hb (342), (e) inhibition of DV cysteine protease activity (264, 318), (f) inhibition of vacuolar phospholipase (343), and (g) inhibition of heme degradation by hydrogen peroxide (H_2O_2) (344). As discussed by Rosenthal and Meshnick (345), the reason that so many mechanisms for CQ have been postulated and that none has easily been proven correct is that CQ may act on a number of targets. It is anticipated that understanding the mechanisms of CQ action will reveal biological details that can be used to circumvent drug resistance by designing a structural analog of CQ effective for malaria chemotherapy (328).

Artemisinin and Analogs

The fast spreading resistance to commonly used quinoline antimalarials has made malaria a major global disease. Among the few alternative drugs available, artemisinin, an endoperoxide antimalarial drug, is the most promising (21). This compound and its derivatives have a broad spectrum of activity throughout the phases of the asexual cycle of the parasite and suppress gametocyte transmission. They decrease parasitemia more rapidly than any other known antimalarial, and are effective against multidrug-resistant parasites (346). Artemisinin is the active ingredient of an ancient Chinese herbal remedy, Qinghaosu (*Artemisia annua*), which was traditionally employed for treating fevers (347). Several semisynthetic analogues of artemisinin have been increasingly used in the past two decades. These compounds encapsulate the active trioxane pharmacophore and include artemether, arteether and artesunate, which are all metabolized to dihydroartemisinin, the main active agent in the body (346). From the beginning, it was proposed that their mode of action was entirely different from those of traditional quinoline antimalarials (348), raising the possibility of finding new antiparasitic targets. In spite of intensive research, the mechanism of action of artemisinin remains a matter of debate (21).

Artemisinin and derivatives are thought to exert their lethal effects either by reductive cleavage of their endoperoxide bond to generate oxygen-centered radicals, or through peroxide ring-opening via protonation or formation of a complex with a metal ion to generate an open hydroperoxide or metal peroxide. Each bioactivation pathway would lead to different alkylated or hydroxylated products on reaction with target molecules, causing free radical-induced damage to the parasite; however, there is no consensus about how free radicals are generated and how the parasite dies (346). It is believed that this process most probably occurs through interaction of artemisinin with intraparasitic free iron or with heme. Indeed, artemisinin taken up by malaria parasites growing *in vitro* is selectively concentrated in the DV and is associated with HZ. Similarly, artemisinin was found to interact with heme, forming covalent adducts, which may be the mechanism of its accumulation and toxicity in the parasite DV (349). In contrast, it was also reported that the artemisinin-heme complex does not possess any antimalarial activity. This may be because several changes occur during heme-artemisinin complex formation, and a fraction belonging to either of the components may be lost during the formation of the complex *in vitro* (350). A two-step mechanism proposed by Metchnick *et al.* (351) suggested that the heme-catalyzed cleavage of the endoperoxide bridge forms a free radical followed by specific and selective alkylation of some plasmodial proteins, and subsequent death of the parasite. The effects of artemisinin and analogues on membrane proteins of parasite vacuolar networks are indicative of possible targets (352).

In parallel, Pandey and colleagues (350) investigated whether this endoperoxide acted by affecting HZ synthesis and whether the presence of preformed HZ was necessary for its antimalarial action. In presence of artemisinin, inhibition of vacuolar proteolytic activity was detected *in vitro* as well as Hb accumulation *ex vivo*. In addition, artemisinin interfered with HZ formation mediated by plasmodial HRP II or parasite lysates, and caused a concentration-dependent breakdown of HZ. Thus, the authors postulated that artemisinin and its derivatives exerted their effect, at least in part, by interfering with the plasmodial Hb catabolic pathway and blockage of heme crystallization. In contradiction with this notion, it was reported that artemisinin antimalarials do not inhibit β-hematin formation *in vitro* (353). These observations are in agreement with an earlier study in which artemisinin-heme adducts did not cause a reduction in HZ content in living parasites (354), and are consistent with the current hypothesis that heme is the activator of artemisinin but not its target (355).

Alternatively, various groups support a non-heme iron activation process for artemisinin (Reviewed by Wu (348)). Iron chelators were found to antagonize the antimalarial effect of artemisinin (356), suggesting that exogenous iron is required for bioactivation because the iron chelators would be unable to sequester iron from ferrous heme (346). It further follows that interactions of the artemisinin antimalarials with heme and free iron seem to represent competitive decomposition pathways, and thus, if it were possible to design derivatives in which degradation of the peroxide bond by heme or free iron is inhibited, highly active drugs would be obtained (353). More work is indeed required to acquire a satisfactory understanding of the antimalarial activity of this class of compounds. Nevertheless, the current evidence suggests that a specific, as yet unknown target is involved (346). In addition, given that artemisinin and its analogues have short half-life and might exert some neurotoxic activity (357), research chemists are attempting to identify semisynthetic and fully synthetic endoperoxides that will retain the activity of the artemisinin derivatives, but overcome some of these deficiencies (21).

Non-iron porphyrins

Non-iron porphyrins PPIX and hematoporphyrin have been shown to bind heme and to interfere with the process of β-hematin formation at acidic pH (358). It is postulated that the interaction between these porphyrins and heme leads to the formation of π-π adducts, which inhibits crystal elongation in the same way as antimalarial drugs such as CQ. Additionally, in the case of hematoporphyrin, the presence of hydroxyl groups able to bind heme iron appears to enhance its activity, as it occurs with the antimalarial drugs quinine and quinidine (359). Consistent with this, Martiney and colleagues (360) showed that various heme analogs (tin PPIX, zinc PPIX, zinc deuteroporphyrin IX and 2, 4 bisglycol) inhibited HZ formation in *P. falciparum* trophozoite extracts. However, all the PP tested failed to inhibit *P. falciparum* growth, which is thought to be due to the inability of these compounds to gain access to the iRBC membrane and to the parasite DV. In contrast, in thalassemic RBC, erythrocytic zinc PPIX seems to gain access to the DV since this compound was found associated with heme (361).

In line with these earlier reports, a subsequent study by Iyer and colleagues (362) demonstrated that zinc PPIX binds to heme crystals and inhibits both crystal formation and extension *in vitro*. Its mode of action was found to be similar to the antimalarial quinolines, binding heme in a saturable, specific, pH and time-dependent manner. Erythrocytic zinc PPIX can elevate significantly in some of the anemias (*e.g.*, β-thalassemia trait) associated with malaria disease protection (363). Thus, these data, along with those of Martiney and colleagues (360), suggest that the mechanism for such protective effect might reside on the ability of this porphyrin to bind heme and block further crystallization. Altogether, these studies show the relevance to design non-heme analogs capable to enter the parasite DV at to coordinate with heme to prevent HZ formation. Alternatively, a better delivery system, using lyophilized liposome formulations of these heme analogs could circumvent the blockage of their entry to the DV (361).

Endogenous Ferrous Heme

In the presence of molecular oxygen in the parasite DV, Fe^{II} PPIX is oxidized to form Fe^{III} PPIX, which is, in turn, sequestered into HZ crystals (265). Reduced iron (Fe^{II}) does not allow the formation of the carboxylate bond needed to obtain HZ (246, 282), but given its electronic structure supports the formation of π-π interactions. Thus, it was hypothesized that Fe^{II} PPIX could inhibit HZ formation. Indeed, Monti and colleagues (364) demonstrated that Fe^{II} PPIX was an inhibitor of β-hematin formation as effective as non-iron porphyrins (namely hematoporphyrin), and was more potent than CQ as measured by molar equivalents. These results point to a fundamental requirement for malaria parasites to survive: the need for complete oxidation of Fe^{II} PPIX to Fe^{III} PPIX to avoid inhibition of HZ synthesis. In line with these observations, β-hematin formation was also prevented by reducing agents such as thiol-containing compounds, corroborating earlier reports (295).

As mentioned above, GSH has been implicated in the degradation of monomeric heme in iRBC, a reaction requiring the presence of oxygen and Fe^{III} PPIX and that is likely to occur *via* the generation of Fe^{II} PPIX (337). Therefore, current research focuses on reducing agents that will be more appropriate and effective antimalarials; in other words, compounds that: (i) are not oxygen scavengers, since this would favor the parasite to counteract the ROS generated during HZ synthesis; and that (ii) are not susceptible to iron-catalyzed oxidation of the functional groups, which would lower the availability of the reducing agent (364). Of interest, small amounts of Fe^{II} PPIX are necessary for the activation of the peroxide bridge of artemisinin-type molecules (365). Thus, compounds that reduce Fe^{III} to Fe^{II} could synergize with these antimalarial endoperoxides. Further application of such compounds alone as well as in combined therapy with artemisinin derivatives is envisaged (364).

Ferric Iron Chelators

Iron is an essential micronutrient required by the malaria parasite for many biochemical reactions involved in growth and multiplication (366). Although the intraerythrocytic parasite lives in an iron-rich environment, ferrous heme resultant from Hb digestion is oxidized and is then detoxified to HZ. Thus, it is most likely that the parasite obtains its iron exogenously (*e.g.*, serum iron, free RBC iron) (324). The importance of the uptake of extracellular iron has led to the use of iron chelating agents as antimalarial drugs. The best studied is the ferric iron chelator is desferrioxamine (DFX). Its antiparasitic activity has been demonstrated *in vitro*, *in vivo* and in patients with both moderate and severe *P. falciparum* malaria (Reviewed by Gordeuk *et al.* (367)).

DFX was found to exert its cytocidal effects only in HZ-containing parasites (trophozoites and early schizonts) (368), raising the question as to whether DFX was acting not only by chelating iron from the parasitic compartment but also from HZ. In this regard, Van Zyl and colleagues (369) showed that DFX had a potent chelating action directly against HZ at the pH of the DV. In contrast, the ferrous chelator 2,2'-bipyridyl was unable to affect the HZ crystal illustrating that the iron associated with HZ is ferric in nature. Thus, the possible disruption of the HZ crystal by DFX would release toxic Fe^{III} PPIX, which is lytic to the parasite. The excess of ferric iron (Fe^{III}) released from the crystal would result in the production of ROS (*e.g.*, H_2O_2, superoxide anion [O_2^-]), which in turn promote membrane damage via lipid peroxidation, enzyme and DNA damage (370). It has been observed that there is an additive effect when DFX and CQ are combined *in vitro* and in clinical trials (371), which may be explained by the fact that both antimalarials react differently on HZ. Whereas CQ might prevent the formation of HZ, DFX may have a direct effect on the structure (369). Thus, a combined therapy of DFX and other powerful antimalarial drugs might help to overcome some of the disadvantages of this iron chelator, namely its relative high cost and the requirement for administration by continuous infusion (324).

Cytochrome P450 inhibitors

Nitrogen heterocycles ellipticine and various azole antigfungal drugs (clotrimazole, econazole, ketoconazole and miconazole) are cytochrome P450 (CYP) inhibitors that bind heme. Some of these antimycotics have the ability to inhibit the *in vitro* growth of *P. falciparum* (372). Of interest, ketoconazole, miconazole and clotrimazole were recently shown to reversibly block HZ crystal growth (336). Thus, beyond their putative antimalarial mechanisms of inhibition of CYP activity (310) and prevention of GSH-mediated heme degradation (373), their inhibitory effects on HZ crystal formation should also be considered. The abundance of compounds that inhibit CYP by binding heme opens the possibility that currently approved inhibitors might be of interest in developing new antimalarial drugs that block HZ crystal growth by a surface binding mechanism, and possibly interfere with other heme-sensitive *Plasmodium* pathways (336).

Xanthones

The 2, 3, 4, 5, 6-pentahydroxyxanthone (X5) has been identified as a potent antimalarial drug with equal activity against multidrug-resistant strains of *P. falciparum* (374). Analysis of the molecular mechanism of action of X5 and selected analogs suggests that they exert their antiparasitic effect by inhibiting HZ synthesis. These compounds formed soluble complexes with heme and prevented *in vitro* formation of HZ in mildly acidic phosphate solution (375). Based on these observations the authors proposed a model for docking of X6, an X5 analogue, displaying several significant interactions with heme. This model predicts that chemical modifications at specific positions, which improve association with heme carboxylate groups, will result in even greater antimalarial activity. Xanthones congeners containing positively charged groups at these positions are being prepared to test this model (375).

Quaternary Ammoniums

Quaternary and bis-quaternary ammonium compounds have received attention due to their potent antimalarial activity both *in vitro* (376) and *in vivo* (377). They have been rationally synthesized to mimic choline structure, and their lethal effect is based on their ability to block *de novo* phosphatidylcholine synthesis, which leads to the inhibition of membrane biogenesis during the intraerythrocytic stage of the parasite (377). In addition to this mechanism of action, a recent work by Biagini and colleagues (378) indicated that heme binding significantly contributes to the antimalarial effect of these drugs. T16 (1, 12-dodecanemethylene bis [4-methyl-5-ethylthiazolium] diodide), a novel bis-quaternary ammonium compound, exhibited potent antimalarial activity. T16 accumulated (~50%) in the parasite DV and was able to bind heme *in vitro*. This interaction appeared to be critical, since decreasing the parasite heme pool caused a marked antagonism of the antiparasitic effects of T16. In addition, since T16 was found to bind HZ *in vitro*, it is plausible that once T16 binds heme in the parasite, it prevents further crystal elongation by incorporation of the complex in the forming chain. Confirmation of this phenomenon will require a more detailed chemical analysis of the complex. It is important to note that bis-quaternary compounds have not shown any cross-resistance with CQ (377). Although heme binding contributes to T16 antimalarial activity, CQ resistance mechanisms, which essentially affect access to heme by modulating the transmembrane transport of CQ (379), do not affect T16 activity (377). Thus, in addition to their ability to block phosphatidylcholine metabolism, these potent antimalarial drugs appear to exert their effects through another mechanism involving heme binding in the parasite DV.

Nine-Anilinoacridine Compounds

9-anilinoacridines are potent antimalarials that inhibit parasite DNA topoisomerase II both *in vitro* and *in situ* (380). From a series of analogs of 9-anilinoacridine, some were found to complex with heme and to inhibit β-hematin formation *in vitro*. Moreover, the drugs enhanced heme-induced RBC lysis at low micromolar concentrations (381). These results open the possibility for development of 9-anilinoacridine derivatives that target two different functions within the malaria parasite, namely DNA topoisomerase II and β-hematin formation, which should help delay the rapid onset of resistance to drugs acting at only a single site.

Methylene Blue

The antimalarial action of methylene blue (MB) was first noted in the late 19th century, but it was not widely used because of its toxicity (382). Derivatization led to the synthesis of primaquine, which is currently used for treatment of *P. vivax* and *P. ovale* hypnozoites (265). A series of MB azure analogues cured murine malaria caused by *P. vinkei* and *P. yoelii*. Moreover, they complexed with heme *in vitro*, which correlated with their antiparasitic effect on *P. falciparum* growth in culture (383). It is therefore suggested that MB and its congeners act by inhibiting heme crystallization, acting similarly to quinoline antimalarials. Recently, MB and other antimalarial dyes, such as Azure A, were shown to exert an inhibitory effect on HZ crystal growth (336). Even though, the potential toxicity of these compounds may restrict their clinical use, their impressive antimalarial activities suggest that their structure could serve as a lead compound for further drug development (383).

HEMOZOIN ACCUMULATION AND DISEASE DIAGNOSTIC - PROGNOSIS

HZ-containing parasites as well as free HZ are released as the iRBC bursts, and are avidly phagocytosed by circulating monocytes, Nϕ and tissue Mϕ (61, 384-389). By reference to standard hematological measurements, it has been estimated that as much as 200 micromoles of HZ (~3 micromoles/Kg) are released into the circulation of a 70 kg *Pf*-infected human patient who has a 1% synchronized parasitemia (61). Since Laveran's original description that pigment granules were visible not only in the blood but also in leukocytes (235), the presence of HZ in these cells has been a diagnostically useful marker for malaria (390-392).

Later, intraleukocytic pigment was also associated with cellular activation and disease severity. In a study performed in adult patients with severe *P. falciparum* malaria, those who succumbed to infection had significantly higher proportions of HZ-containing Nϕ and Mϕ on admission than did survivors (386). In addition, the count of peripheral HZ-containing Nϕ was a better indicator of bad prognosis than the peripheral parasite count in both adults (386) and children (388). In contrast, in hyperendemic areas, pigmented Nϕ or monocytes did not correlate with outcome of infection (387). Evidence from autopsy and histological studies indicated that HZ accumulates in the reticuloendothelial system, turning the liver and spleen black in cases of chronic or repeated infection (393), where pigment persisted for 6 to 12 months after malaria parasite clearance (394). As patient conditions deteriorated and CM occurred, brain capillaries were also filled with iRBC and HZ-laden monocytes (395). Similarly, phagocytes in other locations, such as the eyes, lungs, and kidneys also appear to be loaded with HZ (396, 397).

In line with these observations, HZ has been detected in liver and spleen of infected mice for at least 6-9 months after malaria parasitemia becomes undetectable (398), and its accumulation has been associated with increased parasite burden and disease chronicity (399). Moreover, histological analysis indicated that HZ accumulation occurred predominantly in the tissue Mϕ (splenic Mϕ and Kupffer cells) of the infected animals (61, 197, 399). In a murine CM model, the amount of HZ in the brain was shown to be higher than in mice that did not develop this complication, indicating that the levels of HZ in this organ may be associated with the cerebral pathology (399). Altogether, these different lines of evidence suggested a clinical correlation between disease severity and HZ phagocytosis and tissue accumulation. However, it was not clear whether this association was only circumstantial or if it was due to a direct effect of HZ on the host immune system. The current available data indicate that HZ exerts immunomodulatory effects; however, the nature of the interaction between HZ and phagocytes, as well as its contribution to the pathogenesis of malaria, and particularly its severe complications, are still controversial. Indeed, both inhibitory (67, 119-122, 196, 197, 199-204) and activating (61, 65, 69, 180, 194, 196, 198, 205) effects have been described in pigment-fed phagocytes.

Immunomodulatory Properties of Hemozoin

Inhibition of Immune Cell Functions

Incubation of *P. falciparum* HZ with either murine Mϕ (61) or human monocytes (389) led to rapid phagocytosis of the pigment, which remained inside these cells for at least 72 h. Although HZ-laden phagocytes retained their ability to degrade heme, HZ remained unmodified (389). Moreover, a confocal microscopy study revealed that the inability of the Mϕ to digest HZ does not derive from impairment in phagolysosome formation or acidification, since both activities were normal even after 72 h of HZ phagocytosis (400). Because HZ is a powerful generator of ROS (67) and of highly potent products of lipoperoxidation (200), it has been postulated that the persistence of HZ in long-lived cells such as Mϕ may thus represent a lasting source of noxious products (389). Although HZ was previously considered an inert storage form of toxic heme (247), several studies have shown that the malaria pigment can cause alterations of certain important functions of the phagocytes (67, 119-122, 196, 197, 199-204).

Following phagocytosis of either HZ or RBC infected with mature stages of *P. falciparum*, monocytes lost their ability to kill ingested bacteria, fungi and tumor cells (401). In line with these observations, Schwarzer and colleagues (67) demonstrated that human monocytes fed with trophozoite-parasitized RBC or with isolated HZ were virtually unable to degrade the ingested pigment even after 48 h, to repeat the phagocytic cycle (24-72 h after HZ uptake), and permanently failed to produce oxidative burst after phorbol ester (PMA) stimulation. In parallel, HZ-laden monocytes displayed a long-lasting oxidative burst and their levels of lipid peroxidation increased up to 6-fold at 48 h; an effect that was inhibited by DFX, an iron chelator. Based on these observations, it was suggested that the cause of HZ toxicity resided in the combined effect of pigment-derived iron and oxidative stress, which could account for the peroxidation of membrane lipids and monocyte functional impairment. In a separate study, comparison of the effects of HZ in monocytes and Nϕ revealed that whereas PMA-induced ROS was reduced in HZ-laden monocytes, it remained unaffected in HZ-loaded Nϕ (198), suggesting that the immunomodulatory properties of HZ are cell-type specific.

Subsequently, Schwarzer *et al.* (199) proposed PKC as a potential target for HZ-inducible ROS and iron leading to phagocyte dysfunction. PKC is inhibited by oxidants (402) and is known to play a crucial role in the cellular functions that HZ impairs, namely phagocytosis of opsonized damaged RBC (403) and generation of oxidative burst through the activation of the enzyme NADPH oxydase (NOX) (404). Consistent with this, membrane-associated PKC was irreversibly inactivated after 5 h in HZ-fed human monocytes, even though a transient activation was detected between 30 min an 5 h (199). A plausible explanation for these findings would include HZ-mediated oxidation of cysteine-rich regulatory domains on PKC (405). Alternatively, oxidative modifications with subsequent proteolytic decay could be hypothesized. However, HZ-induced effects on phagocytes could not be solely ascribed to ROS-mediated damage to PKC because it was impossible to restore its activity and membrane translocation using radical scavengers (199). Given that PMA-mediated oxidative burst (which is impaired by HZ) involves PKC-dependent phosphorylation of the cytosolic component of NOX (406), it was next assessed whether inhibition of membrane-associated PKC by HZ was related to NOX inactivation (200). Even though HZ phagocytosis by human monocyte-derived Mϕ resulted in NOX inhibition (already after 2 h), its phosphorylation by PKC was impaired only after 24 h of HZ phagocytosis. Thus, only long-term inactivation of NOX by HZ seems to be dependent on PKC inhibition. Reconstitution experiments indicated that early impairment of NOX activity was due to damage in its cytosolic subunit(s), preventing the translocation to the membrane of a phosphorylated, functionally active NOX. These observations prompted the Schwarzer group to evaluate the role of alternative mediators for HZ toxicity.

Lipoperoxides have been localized in cell membranes of HZ-loaded monocytes (67). These compounds result from the reaction between oxidative radicals and polyunsaturated fatty acids present in membranes. Lipoperoxides are labile, short-lived and therefore, they generate degradation products, such as aldehydes and 4-hydroxynonenal (HNE), in the case of unsaturated fatty acids (407). HNE is a strong alkylating agent, which spontaneously reacts with thiols and amino groups in proteins and purine/pirimidines; thereby blocking the activity of numerous enzymes (408). When human monocytes were fed with isolated HZ, a marked increase in membrane lipoperoxides and HNE levels was detected (maximal at 5 h and significant up to 12 h), and this was accompanied by increased amounts of PKC/HNE adducts (200). A profound inhibition of purified PKC by HNE was also detected at concentrations that seem to be achievable in HZ-loaded monocytes. These data suggest that the previously reported impairment of PKC (199) might be explained by direct interaction with elevated HNE levels. Importantly, the inhibitory effects observed in HZ-fed monocytes may result from the interaction of HNE not only with PKC but also with other enzymes (200).

Given that HZ has been identified as a potent catalyst of lipid peroxidation (67), its ability to oxidize arachidonic acid (AA) -a major component of lipid membranes- was examined (409). In presence of *P. falciparum* HZ, β-hematin or hematin, AA was oxidized to hydroxyeicosatetraenoic acids (HETE)s, mainly 5- 12- and 15-HETE, in the presence of H_2O_2. These compounds are important vasoactive mediators (410), which can alter endothelial cell permeability (411), stimulate RBC adherence to capillary endothelium (412), and induce chemotaxis and chemokinesis (413). Moreover, large amounts of AA are released from cell lipids in response to brain injury and its oxidation products appear to contribute to the cerebral edema (411). Thus, it is conceivable that HZ-mediated AA oxidation to HETEs might be related to some of the pathophysiology associated with severe and CM (*e.g.,* enhanced vascular permeability, edema, sequestration of iRBC).

Further supporting a role for HETEs in malaria pathology, it was recently reported that both trophozoite-iRBC and isolated HZ contained large amounts of hydroxy fatty acids, including the hydroxy-AA isomers 9-, 11- 12- and 15-HETE, which exerted inhibitory effects on monocyte functions (201). Indeed, low micromolar concentrations of HETE isomers were estimated in monocytes after phagocytosis of trophozoites or HZ, and submicromolar concentrations of 12- and 15-HETE inhibited PMA-elicited oxidative burst in monocytes. Of interest, complete delipidization of trophozoites and HZ remarkably reduced their ability to inhibit PMA-inducible oxidative burst in monocytes, suggesting that irreversible inactivation of this phagocyte function was linked to peroxidation of the lipid compartment in parasitized RBC. Thus, in addition to HZ-inducible monocyte lipoperoxidation (67, 414), phagocytosed HZ may ferry large amounts of diverse bioactive lipid derivatives from the iRBC into the phagocytes. Together, both mechanisms might inflict damages to the host even after elimination of the living parasite (201).

Suppression of leukocyte-derived PGE_2 has been associated with enhanced malaria pathogenesis in children (415). Consistent with this, elevated PGE_2 levels in IVBMC were linked to protection against PM in infected pregnant women. Lower levels of phytohemagglutinin (PHA)-promoted PGE_2 were detected in IVBMC in association with increasing amounts of HZ acquired during natural infection (202). These data suggest that *in vivo* acquisition of HZ by placental mononuclear cells may lead to decreased synthesis of PGE_2. Because PGE_2 appears to have a protective role during both CM and PM (202, 415), by reducing its production HZ could contribute to the pathology related to these clinical complications. Even though the mechanism through which HZ exert this inhibitory effect remains to be defined, one potential explanation might reside in the ability of HZ to oxidize AA to HETEs, as described above (409). In vascular tissue, it has been shown that HETEs can suppress PG production (416). Therefore, the possibility that HZ down-regulates PGE_2 biosynthesis via HETEs deserves further investigation.

In addition to its ability to impair monocyte phagocytic and microbicidal functions, a number of studies have reported the down-regulatory effects of HZ on monocyte/Mϕ cytotoxic mediator production in response to proinflammatory stimuli (*e.g.,* bacterial components, cytokines) (121, 196, 197, 203, 204). When murine peritoneal Mϕ were loaded with *P. vinkei* HZ for 24 h and were further stimulated with IFN-γ and/or *E. coli* LPS, the production of IL-6 was drastically reduced after 24 h (196). Similarly, in human endothelial cells, β-hematin inhibited the expression of IL-6 when added before or simultaneously with LPS or TNF-α (203). Moreover, pretreatment with *P. vinkei* HZ for 24 h caused a significant decrease of LPS + IFN-γ-mediated NO production and of zymosan-inducible ROS generation in murine peritoneal Mϕ (197). Consistent with this, β-hematin or hematin (6 to 18 h preconditioning) led to a marked reduction on LPS-inducible NO and TNF-α production in murine peritoneal Mϕ after 48 h of stimulation. In contrast, IL-1 production remained unchanged and that of IL-6 was impaired only at very high concentrations of HZ, indicating a selective inhibition of Mϕ functions. However, the same treatment did not exert any effect in microglial cell lines even if they were able to phagocytose the pigment (121). According to these data, the down-regulatory effects of HZ seem to be due to its heme moiety and may be selective for specific phagocytes and cytotoxic mediators. Addition of antioxidant agents, such as GSH, during the HZ preincubation time partially counterbalanced the inhibitory actions of HZ, suggesting a role for HZ-inducible oxidative stress on peripheral Mϕ unresponsiveness.

This notion was supported by a more recent study showing that HZ inhibited LPS-inducible TNF-α, iNOS and NO production via HZ-mediated ROS generation (204). Free radical production as well as HO and catalase mRNA levels were up-regulated in peritoneal Mϕ following β-hematin treatment. Importantly, thiol compounds partially restored the inhibitory effects whereas iron chelators were unable to do so. These data indicated that HZ-mediated inhibition of Mϕ functions could not be ascribed to iron release from the pigment but rather to HZ-inducible oxidative stress. As to the mechanisms by which ROS might exert these effects, the authors consider several possibilities: (a) depletion of GSH levels required for iNOS induction (417), (b) a direct inhibition of iNOS by H_2O_2 (418), (c) ROS-dependent inhibition of NF-κB-mediated gene regulation (419), and (d) an anti-inflammatory role exerted by HO (420).

In correlation with an earlier study describing the absence of inhibitory effects in microglial cell lines by β-hematin (121), a lower extent of induction of HO and catalase as well as a lower accumulation of oxygen radicals was observed in β-hematin-fed microglial cells than in peritoneal Mϕ (204). Similarly, microglial cells and the Mϕ cell line J774 were found to produce less lipoperoxides in response to β-hematin than peritoneal Mϕ; an effect was attributed to differences lipid composition of their cell membranes and to a higher GSH content (421). Thus, it appears that according to their antioxidant defenses (*e.g.*, GSH levels) and/or their lipid membrane composition, monocytes/Mϕ from different sources may differ in their susceptibility to the pro-oxidant activity of HZ, which in turn determines their responses to the malaria pigment. These findings may help to reconcile reports on both proinflammatory and suppressive activities of HZ in mononuclear phagocytes (204, 421).

One of the major functions of the Mϕ is to assist in initiating and facilitating cell-mediated immune responses by presenting antigenic peptides via surface glycoproteins of the major hystocompatibility complex (MHC) class II (422). In HZ-preconditioned human PBMC (5-24 h), the induction of MHC-II by IFN-γ stimulation (24-36 h) was severely blunted both at the mRNA and protein levels (122). Mechanistically, the abrogation of IFN-γ responsiveness is unexplained. However, since PKC plays a role in IFN-γ-inducible MHC-II expression in Mϕ (423) and PKC inactivation appears to be due to HNE in HZ-loaded monocytes (200), the involvement of HZ-mediated PKC inhibition via HNE might be a potential pathway for MHC-II down-regulation by the malaria pigment (122). The constitutive expression of CD54 (ICAM-1) and CD11c (p150,95 integrin), were also decreased in PBMC after 48 h of *P. falciparum* HZ phagocytosis (122). In human endothelial cells, β-hematin alone did not affect basal levels of ICAM-1; however, when added in combination with LPS or TNF-α, it inhibited the expression of this molecule (203). ICAM-1 contributes to the capacity of APCs to adhere and stimulate T cell proliferation by reinforcing the signal from the T cell receptor and CD11c has been identified as a ligand for ICAM-1 (424). Therefore, down-regulation of their expression by HZ may help to explain the defective T cell response in malaria. In contrast, HZ ingestion had no effect in the constitutive expression of a number of receptors implicated in phagocytosis, including CR1, CR3, CD36, and CD64. Thus, only the expression of molecules that play essential roles in antigen presentation and T-cell response appears to be specifically hindered or blocked after HZ phagocytosis (122). These results suggest that HZ-loading may contribute to the defects in cell-mediated immunity associated with malaria immunosuppression.

In human malaria, immunosuppression appears late in the acute phase of the disease and can last a long time after the clearance of parasites from the circulation (106). Although the precise mechanisms are not clear, early investigations have suggested that defects in Mϕ accessory cell function could account for the poor immune reactivity (425). In *P. berghei*-infected mice, HZ-laden liver and splenic Mϕ displayed accessory cell dysfunction. In addition, either isolated *P. falciparum* HZ or *P. berghei* HZ from livers and spleens of infected animals markedly reduced splenic plaque-forming cell responses to sheep RBC (119). Further supporting a role for HZ in Mϕ accessory cell dysfunction, either *P. chabaudi* HZ or β-hematin were able to mimic *in vitro* the immunosuppressive effects induced in peritoneal Mϕ during *in vivo* infections with *P. chabaudi* (120). Both the native and the synthetic pigment severely affected the ability of Mϕ to activate antigenic-specific T-cell hybridomas to produce IL-2; an effect that was also observed when the peritoneal exudate cells from *P. chabaudi*-infected mice were cocultured with the antigen-specific T-cell hybridomas. These studies suggest that the inhibition of Mϕ accessory cell activity is due, at least in part, to the uptake and accumulation of HZ. However, whether or not similar responses to

HZ occur *in vivo* and the exact cellular mechanisms involved remained to be elucidated. A likely possibility is that HZ interferes with antigen processing and presenting by Mϕ, as suggested by the above-described study indicating down-regulation of MHC-II and adhesion molecule expression (122).

Activation of Immune Responses

Severe malaria infection tends to be associated with increased proinflammatory cytokines in the peripheral circulation (426). Similarly, elevated levels of these mediators have been associated with induction of intermittent fevers during acute malaria, which correlate with waves of schizont rupture and release of parasite components into the bloodstream (95). Several lines of evidence suggest that HZ, a unique schizont component, could play a proinflammatory role during malaria infection. Despite the impairment of some of their functions, HZ-loaded monocytes/Mϕ remain viable and retain their adherence properties for up to 72 h (61, 67, 120). Moreover, these cells preserve normal DNA and protein levels, and retain their respiratory capacity (121, 204), suggesting that rather than a generalized toxic effect HZ may cause different effects on selective targets in phagocytes. In fact, a number of studies have reported the ability of HZ to induce the expression of various leukocyte inflammatory mediators (61, 65, 69, 180, 194, 196, 198, 205).

In 1994, Pichyangkul and colleagues (65) first described HZ as a "potential virulence factor in the monokine-mediated induction of organ-specific and systemic pathology in *P. falciparum* malaria". These conclusions were based on the observation that human PBMC co-cultured with schizont-parasitized RBC or with isolated *P. falciparum* HZ released high levels of TNF-α and IL-1β after a 2- to 8-h treatment. However, protease treatment of the isolated pigment abolished PBMC cytokine production. In addition, β-hematin preparations induced only minimal production of TNF-α and IL-1β, suggesting that the active moiety of HZ was a protein or a protein-linked molecule associated with the malaria pigment. These data are in sharp contrast with those obtained in later investigations in which either β-hematin or native purified HZ was shown to induce monocyte/Mϕ cytokine production (61, 196, 205). Even though the reasons for these inconsistencies remain unclear, they are probably related to differences in the preparation and characteristics of HZ/β-hematin samples used in the various studies (427) as well as in the concentrations of pigment used and the times of incubation. Alternatively, the apparent discrepancies might be also associated with a variable susceptibility of the phagocytes to HZ treatment according to its antioxidant defenses and/or their lipid membrane content, as previously discussed (204, 421).

Both purified *P. falciparum* HZ and chemically synthesized β-hematin potently induced the release of TNF-α, MIP-1α and MIP-1β, from RAW 264.7 murine Mϕ and human PBMC (61). The production of these mediators was concentration-dependent and was sustained, continuing for as long as 72 h. In contrast, phagocytosis of other well-characterized forms of heme, hemin and hematin as well as that of latex beads did not induce the production of any of these proinflammatory mediators, showing the specificity of the stimulus imparted by HZ. Consistent with this, incubation of human monocytes with *P. falciparum* HZ strongly stimulated TNF-α production, an effect that was additively increased by LPS (198). HZ-inducible IL-10 production by monocytes was also apparent, suggesting that HZ might exert both pro- and anti-inflammatory effects in the infected host. In the same study, IL-12 secretion was detected but only in presence of large amount of HZ. In murine peritoneal Mϕ, *P. vinckei* HZ enhanced TNF-α production after costimulation with LPS $\pm$ IFN-γ (196), which might help to explain the sometimes very high TNF-α levels in peripheral circulation of patients with severe malaria and bacteriemia (428). Of interest, Biswas and colleagues (205) reported the production of significant amounts of anti-HZ IgM Abs in malaria patients, with higher levels among complicated *P. falciparum* cases. The anti-inflammatory properties of these Abs were suggested by their inhibitory effect on the *in vitro* production of TNF-α and IL-β by PBMC in response to HZ. Therefore, anti-HZ Abs raised during infection may have a significant role in determining the severity of the disease by controlling the excessive production of cytokines (205).

Severe anemia is one of the main causes of morbidity and mortality related to *P. falciparum* malaria (170). Although the role of host-derived molecules (TNF-α, IFN-γ; IL-1β) as erythropoiesis inhibitors has been explored (179), only IL-12 appears to exert a protective effect (429). In a search for additional host factors contributing to malarial anemia, Martiney *et al.* (180) investigated the implication of MIF. They found that MIF -a proinflammatory mediator and a counter regulator of glucocorticoid action (430)- inhibited erythropoiesis *in vitro* in presence of erythropoietin. In the sera of *P. chabaudi*-infected mice, increased levels of MIF were detected, which correlated with disease severity. Ingestion of *P. chabaudi*-iRBC or HZ induced the release of MIF from peritoneal murine Mϕ. In addition, liver MIF immunoreactivity increased concomitant with extensive HZ and iRBC deposition. Similarly, MIF was elevated in the spleen and bone marrow (organs capable of erythropoietic expansion) at peak levels of parasitemia, where high HZ and iRBC accumulation was observed. These data are in line with ultrastructural analysis of the bone marrow of severely anemic children showing Mϕ containing iRBC and HZ (431), and with a previous report in which HZ-laden Mϕ were seen adjacent to developing erythroblasts in the red pulp of the spleen in infected animals

(179). Altogether these results suggest that HZ-inducible MIF may be a host-derived factor involved in the pathophysiology of malarial anemia.

Another characteristic feature of *P. falciparum* malaria is the cytoadherence of iRBC and leukocytes to vascular endothelium (146). Exposure of Nϕ or Mϕ to isolated *P. falciparum* HZ resulted in significant up-regulation of the adhesion molecule CD11b/CD18 and down-regulation of leukocyte adhesion molecule 1 (LAM-1). Ca^{++}-dependent phagocytosis of HZ appeared to be a pre-requisite for this effect, since EDTA prevented both HZ uptake and leukocyte adhesion molecule modulation (194). The involvement of CD11b/CD18 and LAM-1 in leukocyte adhesion to endothelium during inflammatory responses is well documented (432). Their inverse expression has been detected in parallel upon leukocyte activation (433), including cellular responses to TNF-α, C_{5a} and leukotriene B_4 (LB_4) (434). Thus, it is likely that phagocytosis of HZ leads to release of several mediators, which in turn regulate the expression of these adhesion molecules. Through this mechanism, HZ could promote leukocyte adherence to vascular endothelium *in vivo*, thereby contributing to tissue injury and microvascular flow disturbance observed in malaria.

Further supporting a role for HZ in the activation of the immune response, Coban and colleagues (69) reported the ability of HZ to induce DC maturation *in vitro*. DC, as crucial APCs, are effective in activation of naïve T cells and thus in initiating adaptive immune responses (435). They also activate innate immune responses to microbial challenge by producing cytokines, such as IL-12 (436). When immature human DCs were incubated with either *P. falciparum* or highly purified *P. falciparum* HZ, significant increases in the maturation markers for DCs (CD83, CD86, and CD1a) were detected. In addition, the production of IL-12 and the morphological features characteristic of mature DCs were also observed. In contrast, β-hematin and heme did not alter the expression of any of these cell surface molecules in immature DCs. Since these cells are also important for the *in vivo* immunogenicity of DNA vaccine-encoded antigens (437), the effect of HZ on DNA vaccine-induced Ab responses was also evaluated in the Coban study. Although immunization in presence of HZ did not affect antigen-specific total IgG titers, the ratio of IgG2a to IgG1 Abs was significantly higher in the group injected with HZ. Altogether, these data indicating HZ-enhanced DC maturation and Ab isotype switching during primary immunization warrant further investigation of HZ as a potential immunomodulator (69).

A causal relationship between HZ and proinflammatory mediator release *in vivo* remains elusive. However, a study using a rat model revealed that HZ has the ability to induce changes in body temperature (61), suggesting that this parasite metabolite could contribute to fever during acute malaria, probably *via* the induction of strong pyrogens (*e.g.*, TNF-α, IL-1). In addition, co-localization of HZ and proinflammatory mediator production has been reported in both experimental and clinical malaria (186, 190, 196). *In situ* hybridization in sections of spleens from mice infected with *P. vinckei* showed increased TNF-α mRNA expression in the red pulp surrounding HZ accumulation (196). In human PM, production of MIP-1β and MCP-1 was observed in some of the maternally derived HZ-laden placental Mϕ (186). Similarly, TNF-α and IL-8 were produced only by placental Mϕ that contained HZ. Moreover, higher TNF-α mRNA levels in the placenta were associated with increased HZ content (190). Based on these different lines of evidence as well as on its effects *in vitro*, it is conceivable that HZ exerts immunomodulatory effects on the host. Further studies are required in order to understand the molecular mechanisms involved in HZ-inducible phagocyte activation, and to assess the immunomodulatory properties of HZ *in vivo*.

CHAPTER 3
NITRIC OXIDE

INTRODUCTION

NO is a labile and highly reactive gas with a molecular weight of 30 daltons, which diffuses freely through cell membranes (438). Formerly, NO was mainly considered as a member of undesirable environmental pollutants and potential carcinogens. Interest in the physiological function of NO was potentiated by its identification as the endothelium-derived relaxing factor mediating vascular smooth muscle relaxation and vasodilatation (439) and as a neuronal messenger molecule (440). Since then, NO has been further identified as a major regulatory molecule of the immune system and as a principal cytotoxic mediator of activated immune effector cells (441). The first nonmicrobial species in which production of NO was established was the human, and the first cell type was the murine Mϕ. These findings have now extended to many other organisms, including birds, insects, and protozoa (438). Generation of NO is a feature of immune-system cells (*e.g.,* DC, NK cells, mast cells, monocytes, Mϕ, microglial cells, Kupffer cells, eosinophils and Nϕ) as well as other cells involved in immune reactions, such as endothelial cells, epithelial cells, vascular smooth muscle cells (VSMC), fibroblasts, keratinocytes, chondrocytes, hepatocytes, and mesangial cells (442). In light of the significant role NO plays as a messenger molecule and regulator of cellular functions, it was named "molecule of the year" in 1992 (443).

NITRIC OXIDE SYNTHESIS

The biochemical pathway of NO synthesis in mammalian cells was initially elucidated in activated murine Mϕ. NO is formed from the oxidative deamination of L-arginine to produce L-citrulline. This reaction is catalyzed by the enzyme NO synthase (NOS), which in addition to the substrate L-arginine requires molecular oxygen, NADPH, and other cofactors, such as calmodulin, flavin adenine dinucleotide (FAD), flavin adenine mononucleotide (FMN), heme, and tetrahydrobiopterin (BH_4) to produce NO and citrulline (444).

Three different isoforms of NOS, two constitutive (cNOS) and one inducible (iNOS), have been identified. The cNOS -NOS I in neurons (nNOS) (445) and NOS III in endothelial cells (eNOS) (446)- are involved in homeostatic processes such neurotransmission, peristalsis and blood pressure regulation. These constitutive low-output forms of NOS produce small amounts of NO (on the nanomolar level) for short periods, and are activated by calcium binding to the enzyme cofactor calmodulin (447). The third form of NOS, iNOS or NOS II, is not constitutively expressed but is induced in response to a variety of stimuli, such as cytokines and pathogen components (448) and its expression is tightly controlled both at the transcriptional and post-transcriptional levels. Neuronal NOS was the first of these enzymes to be cloned. The molecule presents binding sites for calmodulin as well as for NADPH, FAD and FMN, which support the oxidation of a guanidine nitrogen of arginine (445). The iNOS protein shares about 50% identity in amino acid sequence with eNOS and nNOS, including the binding sites for multiple cofactors (449). Mϕ iNOS has been reported to contain a tightly bound calmodulin molecule, which maintains an active conformation; thus, in contrast to cNOS, its activity is not regulated by enzyme activation or inhibition depending on calcium binding, but is rather controlled by the rate of enzyme turnover (450). In addition to Mϕ, many other cell types, including fibroblasts, hepatocytes, astrocytes and endothelial cells possess the iNOS enzyme. The levels of NO produced by the iNOS enzyme are far higher than those produced by cNOS, and this is therefore referred to as the high-output pathway of NO synthesis (441).

Both cNOS and iNOS are initially produced as inactive monomers. The NOS molecule has a heme-binding site in its N-terminal domain and contains Fe-PPIX. Dimerization of NOS leads to approximation of the C- and N-terminal domains in a *trans* arrangement, allowing the reduction of the Fe^{III} moiety required for binding molecular oxygen (444). In the case of cNOS enzymes, dimerization follows binding of L-arginine, BH_4 and heme. Binding of calmodulin -which has been activated by high calcium levels- induces a further conformational change in the dimer, resulting in enzymatic activity. In the case of iNOS, calmodulin binds to the molecule even in the presence of only trace amounts of calcium. Dimerization and enzymatic activity occur following the binding of BH_4, L-arginine and heme. As currently understood, the mechanism of action of NOS involves an initial hydroxylation of L-arginine to N^G-hydroxy-arginine followed by an additional oxidation (451).

The function of all forms of NOS can be inhibited by N-substituted arginine analogs such as N^G-monomethyl-L-arginine (NMMA), aminoguanidine, or *N*-nitroarginine methylester. The availability of such inhibitors has facilitated assessment of the effects of the arginine-NO pathway *in vivo* (452). In culture, NO synthesis is also inhibited by the activity of the arginase enzyme, which utilizes arginine to form ornithine and urea, thus depleting the substrate for NOS (453). Flavoprotein binders or calmodulin inhibitors can also inhibit the activity of NOS (454). In addition, NOS is inhibited by NO itself through a direct effect on enzyme-bound heme (455).

NITRIC OXIDE CHEMISTRY

To understand the pathophysiological actions of NO and the paradoxical beneficial versus deleterious effects that NO sometimes exhibits, one must first understand the biological chemistry of NO (438). In biological fluids, NO has a half-life of only a few seconds because is readily oxidized to nitrite (NO_2^-) and subsequently to nitrate (NO_3^-), which are excreted in the urine (456). The effect of NO on cells ultimately depends on many complex conditions such as the rate of NO production and its rate of diffusion; the concentrations of potential reactants, such as O_2 and O_2^-; the levels of detoxifying enzymes, such as catalase and superoxide dismutase (SOD); the levels of antioxidants (*e.g.,* GSH); and the distances between generator cells and target cells (457).

The biochemical basis of the cytostatic and cytotoxic functions of NO appears to be primarily due to DNA damage and the inactivation of critical metabolic pathways (458). Autooxidation of NO forms nitrous anhydride (N_2O_3) and reaction with O_2^- forms peroxynitrite ($ONOO^-$), which rapidly decay to release the highly reactive hydroxyl radicals ($OH^\bullet$) (459). Through these species, NO can cause DNA damage: (i) $ONOO^-$ can oxidize and nitrate DNA and may cause strand breaks through attack on the sugar-phosphate backbone; (ii) N_2O_3 can nitrosate secondary amines to form nitrosamines that, after metabolic activation, can alkylate DNA through an indirect mechanism; (iii) N_2O_3 can also directly nitrosate the primary amine functionalities of DNA bases leading to deamination. $ONOO^-$ can also induce oxidative and nitrative damage to biological components other than DNA, including lipids and proteins. Moreover, both DNA-DNA and DNA-protein cross-links can also be formed through NO-derived reactive species (460).

In addition to reaction with ROS, NO can react with transition metals, thiols and amines (461). Through the formation of iron-dinitrosyl-dithiolate complexes, NO inactivates several metabolic key enzymes with iron-sulfur (4Fe-4S) prostetic groups at their catalytic sites (*e.g.,* enzymes of the Krebs cycle and the electron transport chain); thereby interfering with important biological processes, such as DNA synthesis and cell proliferation (458, 462, 463). The balance between oxidation and nitration is influenced by a number of factors, including the presence of carbon dioxide (CO_2). Nitrosoperoxycarbonate ($ONOOCO_2^-$) is postulated to be the first intermediate in the oxidation and nitration chemistry. The mechanism of tyrosine oxidation and nitration appears to involve an initial oxidation by $ONOOCO_2^-$ to generate intermediate $OH^\bullet$ and $NO_2^\bullet$ radicals (460). In addition to Mϕ reactive species, NO derivatives can react with Nϕ products. The enzyme myeloperoxidase (MPO) has been shown to effect tyrosine nitration when NO_2^- is present as a co-substrate, through a mechanism that seems to be mediated by nitryl chloride (NO_2Cl) -a potent tyrosine nitrating and chlorating agent- and $NO_2^{\bullet}$ produced during MPO-dependent oxidation of NO_2^- (464). Interactions between ROS and RNI generated simultaneously by phagocytic cells provide a molecular basis for a more cytotoxic combinatory effect of the respiratory burst and NO synthesis on target cells.

NITRIC OXIDE ACTIONS

Homeostatic Functions

Owing to its small size and exceptional membrane permeability, NO is well suited to serve as a chemical messenger in cell-cell interactions. NO and its derivatives are involved in intercellular communications regulating blood vessel dilation and postsynaptic-to-presynaptic neuronal interaction (465). The function of the relatively small amount of NO generated through the low-output neuronal and endothelial cNOS mechanisms appears to center around its ability to stimulate the enzyme guanylate cyclase to produce elevated levels of cyclic GMP (cGMP), probably through NO binding of heme iron in the active site of the latter enzyme. Intracellular GMP then serves as a second messenger in the activation of protein kinases (466).

In the cardiovascular system, the release of NO acts as a general adaptive mechanism whereby the vascular endothelium responds to changes in its environment and regulates blood flow and blood pressure through an action on the vascular smooth muscle. In the nervous system, the importance of the NO pathway is linked to the stimulation of specific receptors in the central nervous system. Both nNOS and guanylate cyclase are widely but not uniformly distributed in the brain. Thus, it is likely that this pathway is associated with other mediator systems (467).

Nitric Oxide and the Immune System

In recent years, NO has been found to play a much more diverse role in infection and immunity than it was initially assigned. Its numerous actions include antimicrobial and antitumor activities as well as immunosuppressive, tissue-damaging and pro/anti-inflammatory effects. The ideas that NO is always produced at high levels in the immune system, is derived form iNOS, and has host-protective effects during infection and tissue-damaging effects during autoimmune responses are oversimplifications (442). It is now clear that iNOS is detrimental in some infectious processes (441), it helps to counteract excessive immune reactions (468), protects to some degree against autoimmunity (469), and functions as intra- and intercellular signaling molecule shaping the immune response (470). In addition, nNOS and eNOS are now known to participate in important immunological processes such as chemotaxis and apoptosis (471, 472). Importantly, the demonstration of iNOS expression by Mϕ and other cell types in tissues from patients with a wide variety of infectious, autoimmune and degenerative diseases has disproved the claim that iNOS does not occur in the human immune system (442).

Antimicrobial Effects

A particularly important role of NO is its involvement in host defense. The antimicrobial activity of NO has been demonstrated by a variety of approaches. First, there is circumstantial evidence implicating NO production as part of an effective host response to infection. NO generated by iNOS is stimulated by proinflammatory cytokines such as IFN-γ, TNF-α, IL-1 and IL-2, as well as by microbial products such as LPS and lipoteichoic acid (473). Infections in humans and experimental models are often associated with significant increases in systemic NO production -when determined by measurement of NO end-oxidation products in plasma and urine (143, 474)- and with increased iNOS expression and NO production at sites of infection (*e.g.,* leishmaniasis, tuberculosis) (475, 476). Second, production of NO in many animal models can be directly correlated with the ability of the host to restrain microbial proliferation, because abrogation of iNOS activity produces dramatic increases in microbial burden (452, 476, 477). Third, *in vitro* studies of a variety of cells (*e.g.,* Mϕ, Nϕ, hepatocytes, endothelial cells) acting in a diverse array of microbial targets have demonstrated pathogen- and/or cytokine-inducible microbiostatic or microbicidal activity, which is L-arginine-dependent and is blocked by competitive NOS inhibitors (441, 473). Finally, NO donor compounds have been shown to inhibit or kill microorganisms when directly administered *in vitro* (478). Although susceptibility is not universal, NO-related antimicrobial activity has been reported against a remarkably broad range of pathogens including viruses (*e.g.,* Epstein-Barr virus, murine cytomegalovirus [CMV], bacteria (*e.g., Salmonella typhimurium, Mycobacterium tuberculosis* [*M. tuberculosis*]), and parasites (*e.g., P. falciparum, L. donovani, L. major, T. cruzi, T. gondii*) (441, 442, 473).

In murine Mϕ, it has been demonstrated that NO is produced as a cytotoxic agent in the immune or inflammatory response (456, 479). Even though NO is released from Mϕ in combination with other species including H_2O_2 and proinflammatory cytokines, it appears to be the key mediator of Mϕ-induced cytostasis because NO scavengers block this effect (480). In contrast to murine Mϕ, the role of NO-related antimicrobial activity in human mononuclear phagocytes has been one of the most controversial issues in NO biology. Even though it is clear that NO production is elevated in infected patients or in recipients of proinflammatory cytokines (*e.g.,* IL-2 administration) (458), it is possible that this NO originates in cells other than Mϕ (473). Some investigators have found that human Mϕ from normal subjects produce little or no NO in response to stimuli that elicit high quantities of NO from Mϕ of murine origin (481, 482). In contrast, others have reported low levels of L-arginine-dependent NO production in response to various stimuli, including HIV-1 infection and CD23 or CD69 ligation (483-485). Importantly, several independent studies have demonstrated induction of iNOS mRNA and protein in human Mϕ. The most compelling evidence has been obtained from patients with inflammatory conditions. Elevation in iNOS protein expression by circulating mononuclear cells has been correlated with increased systemic NO production in Tanzanian children with malaria (143). Similarly, the iNOS protein has been detected in alveolar Mϕ from patients with active pulmonary tuberculosis, in the cerebral cortex of AIDS patients with severe dementia, in PBMC of patients with hepatitis C and rheumatoid arthritis, and in the skin of patients with tuberculoid leprosy or localized cutaneous leishmaniasis (475, 486-489).

Further supporting an important role for human monocyte-derived NO production, in a study of the effect of IFN-γ-treated human monocytes on asexual blood forms of *P. falciparum*, incubation in presence of NOS inhibitors partially reduced their antiparasitic activity (490). In line with these data, it was reported that pretreatment with IFN-γ plus TNF-α activated human Mϕ for trypanocidal activity through a NO-dependent mechanism (491). Taken together, these different lines of evidence suggest that high-output NO synthesis is indeed part of the antimicrobial actions of human Mϕ, but the signals required for stimulation of this system remain incompletely understood (473). As stated by James (441), it is possible that human monocytes must receive additional signal(s) before becoming fully capable of producing NO. Identification of some missing element in the production of NO by these cells is required to fully determine the contribution of this molecule to human immunity.

Pathological Effects

NO formation may have originated as a first-line defense against invading microbial organisms; however, the cytostatic/cytotoxic effects of NO might be detrimental not only to the pathogen but also to the host. In several inflammatory and autoimmune conditions (*e.g.,* arthritis, asthma, diabetes, glomerulonephritis) as well as in various infectious diseases (*e.g.,* CM, HIV-1 infection, tuberculosis), the expression of iNOS is clearly associated with a more severe or even a fatal outcome (441, 492, 493). Possible underlying mechanisms include NO-mediated cytotoxicity and tissue damage, inhibition of T cell proliferation and/or induction of T cell apoptosis, and direct positive effects of pathogen growth (442).

A number of studies on the role of NO in the immune response have revealed that -depending on the species, strain, infection dose and pathogen entry route- iNOS is indispensable or helpful in the control of infection (*e.g., T. cruzi, L. donovani*) whereas in other cases, it does not exhibit any discernable effect (*e.g., Streptococcus pneumoniae* [*S. pneumoniae*], *T. brucei*) or it exacerbates the disease (*e.g.,* Influenza virus, *T. gondii*) (reviewed by Bogdan (442)). A key concept in understanding the dual role of NO in physiology and pathophysiology is that the local concentration of NO synthesis is an essential determinant of cytotoxicity: picomolar amounts of NO are sufficient for intracellular signaling while micromolar concentrations are microbicidal as well as proinflammatory and damaging to the surrounding cells and tissues (494). Other factors that should be taken into account when evaluating the effects of NO are the timing of its production and the biological milieu in which it is released (441). In several disease models, the antimicrobial versus detrimental effects or the lack of influence of NO is restricted to certain organs and/or stages of the infection. For instance, during infections with *T. gondii*, iNOS enhances (intestine) or inhibits (central nervous system) the severity of the disease; and in infections with *T. cruzi*, iNOS is required to control parasite burden during the acute but not in the latent phase of the disease (494, 495).

Because most microbial agents induce a period of inflammation in the infected host, NO production is a common finding during numerous pathological conditions. The length of NO production and surrounding conditions differ according to the type of infection. Thus, it is the balance between NO and other cytokines that is crucial in determining the effects of NO in the host (441). To add further complexity, it is now clear that host cells are differentially susceptible to NO. For example, while hepatocytes are quite resistant, cells in the gut are extremely susceptible to damage induced by high levels of NO (496, 497). Hence, the target organs of invading pathogens may differ in their susceptibility or resistance to NO toxicity (441). Figure 1 illustrates some of the multiple factors that can influence the production and effects of NO during infection, thereby contributing to the positive or negative outcome of a given disease (498).

MECHANISMS OF REGULATION OF iNOS EXPRESSION

The expression of iNOS is regulated primarily by *the novo* synthesis and stability of iNOS mRNA and protein (499). The iNOS isoform is positively or negatively regulated by cytokines (*e.g.,* IFN-γ, TNF-α, IL-1, TGF-β, IL-4, IL-10); pathogen products (*e.g.,* proteins, lipids, polysaccharides); cell-cell contact (via adhesion and costimulatory molecules), immune complexes, polyamines, non-ferritin bound iron, oxygen tension, environmental pH, and various antibiotics (438, 500). Although

IFN-γ and LPS are the best-studied examples, novel regulators continue to be discovered. For instance, IL-12 (with IL-18) induces iNOS in various populations of Mϕ, through a mechanism mediated by autocrine production of IFN-γ (501).

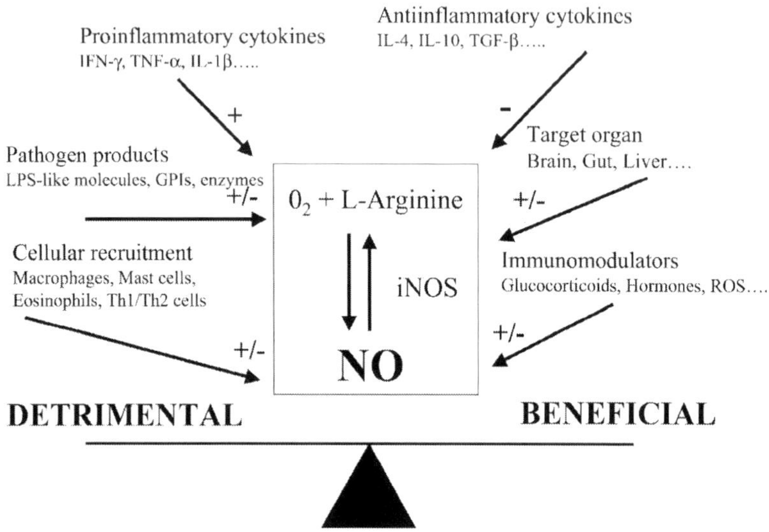

Figure 1. Potential Factors that Influence the Role of Nitric Oxide Production and its Effects during Infection. Some factors (depicted as +) favor NO production, whereas others (depicted as -) down-regulate NO production. Factors depicted as $\pm$ have no *a priori* effects on NO synthesis, and their effects depend on the surrounding milieu. The balance between detrimental and protective effects of NO will be altered by production of hormones, glucocorticoids, ROS and RNI, and by pathogen-induced cell recruitment. In addition, cells in different organs are differentially susceptible to NO. Finally, pathogen products themselves may directly induce NO production or inhibit its activity. Reprinted from *International Immunopharmacology*, Vol 1:1457-1467, Brunet, L. R, Nitric oxide in parasitic infections, Copyright 2001, with permission from Elsevier. Reference (498).

Microbial products that have been shown to stimulate NO production by Mϕ include the human T lymphotropic virus type 1 (HTLV-1) transactivator Tax, the 19-Kd lipoprotein of *M. tuberculosis* (acting *via* TLR2), the flagellin of Gram-negative bacteria (acting *via* TLR5), the GPI of *P. falciparum*, bacterial DNA and CpG-containing oligodeoxynucleotides (acting *via* TLR9), and DNA from various protozoan parasites (*e.g., Babesia bovis, T. cruzi, T. brucei*) (66, 502-505).

Transcriptional Control

Activation of the *iNOS* gene promoter is an important mode of iNOS regulation, which has been analyzed most thoroughly in murine Mϕ and in human hepatocyte and epithelial cell lines. The participating transcription factors include NF-κB, activator protein-1 (AP-1), signal transducers and activators of transcription (STAT)1α, IFN-γ responsive factor-1 (IRF-1), and CCAAT/enhancer binding protein (C/EBP)β (506-508). Depending on the cytokine or microbial stimulus and the cell type, different upstream signaling pathways are involved, which either promote or inhibit iNOS expression. For instance, Janus kinases (Jak): Jak1, Jak2 and Tyk2; mitogen-activated protein kinases (MAPK): p38, extracellular signal-regulated kinase (ERK)1/2 and c-Jun amino-terminal kinase (JNK); PKC; Raf-1 protein kinase; cyclic adenosine 5' -monphosphate (cAMP); and protein phosphatases 1 and 2A up-regulate iNOS expression (509-516). In contrast, other proteins, including phosphatidylinositol 3-kinase (PI-3-K) (517, 518) and protein tyrosine phosphatases (PTP) (517) are able to down-regulate iNOS synthesis. In addition, NO itself exerts a biphasic effect on iNOS transcription. Low concentrations of NO (such as occur at the onset of Mϕ stimulation by cytokines) activate NF-κB and upregulate iNOS (positive feedback) (519). In contrast, high concentrations of NO have the opposite effect, which may help to prevent NO overproduction (520).

The cloning of the murine *iNOS* promoter was accompanied by the demonstration that the regulation of iNOS induction by IFN-γ and LPS is transcriptional (450), although there may be an additional effect on mRNA stability (438). Transcriptional control of iNOS occurs *via* a complex structure of promoter proximal and more distant enhancer elements. Nearly twenty-four transcription factor binding-site elements exist within the *iNOS* promoter (438). The array of putative transcription factor recognition boxes include: three IFN-γ-activated sites (GAS), two NF-κB sites, two IFN-stimulated response elements (ISRE)s, two recognition sites for AP-1, two TNF-response elements (TNF-RE), and one X-box (521).

Functional dissection of the murine *iNOS* promoter focused first on responses to LPS in the RAW 264.7 Mϕ cell line and implicated the transcription factor NF-κB (522). In this study, it was shown that LPS leads to the assembly of mixed heterodimers of c-Rel and p65 with p50 in the downstream NF-κB site, which is located in an essential *cis*-acting element in the murine iNOS promoter. These findings were subsequently extended to rat alveolar Mϕ (523) as well as to epithelial cells and VSMC in rats and mice (438). Further analysis of the elements controlling the expression of the murine *iNOS* gene revealed that its enhancer region contains an upstream NF-κB site that -in contrast to the downstream site- is not essential for *iNOS* expression in response to LPS $\pm$ IFN-γ, but which rather appears to function as an enhancer. Specific mutations in this site reduced the expression of the iNOS promoter in response to LPS and abolished up-regulatory effect of LPS in IFN-γ-mediated *iNOS* induction. These data indicate that the NF-κB upstream site plays an important role in maximal induction of the *iNOS* gene by LPS alone or in combination with IFN-γ (524). Although both NF-κB sites bind the same heterodimers (p50/c-Rel and p50/p65) (522, 525), they seem to have different functions. Thus, it is likely that their positions in the iNOS promoter and the interaction of NF-κB bound to each site with other transcription factors give different function (524). In fact, according to the current data, the major contribution to synergistic iNOS promoter activation appears to be the cooperative action of NF-κB with STAT1α and/or IRF-1 (526-529).

The enhancer region of the *iNOS* gene also contains an IRF element (IRF-E) and a GAS, which are essential for *iNOS* expression (530-532). Mice rendered genetically deficient in the transcription factor IRF-1 were grossly deficient in expression of iNOS mRNA and NO production in response to IFN-γ + LPS (530). Moreover, mutational, binding, serologic, and cross-linking analyses demonstrated that the contribution of IFN-γ to the synergistic activation of *iNOS* by IFN-γ + LPS required IFN-γ-induced binding of IRF-1 to the IRF-E in the iNOS promoter (531). GAS elements are known to bind homodimers of a phosphorylated form of the transcription factor STAT1α, which is a latent cytoplasmic protein. Treatment of cells with IFN-γ causes tyrosine phosphorylation of STAT1α by the IFN-γ receptor-associated kinases Jak1 and Jak2. Subsequently, phosphorylated STAT1α forms homodimers and translocates into the nucleus where it induces transcription of GAS-containing genes (533). The GAS element located in the enhancer region of the murine iNOS promoter was found to be required for optimal transcriptional activation in response to IFN-γ, LPS and IFN-γ + LPS. Furthermore, STAT1α -unlike STATs 3 to 6- was shown to bind *iNOS* GAS in IFN-γ-, LPS-, or IFN-γ + LPS-stimulated murine Mϕ, providing evidence for the direct involvement of STAT1α in the transcriptional activation of the murine *iNOS* gene (532). In addition to having a direct role in *iNOS* activation, STAT1α can also act indirectly since it is required for induction of the *IRF-1* gene (534). Targeted disruption of the *STAT1α* gene abolished Mϕ NO production in response to IFN-γ + LPS (535); however, it cannot be concluded from these data whether the effect is attributable directly to the absence of STAT1α, indirectly to a relative lack of IRF-1, or both.

Taken together, these studies illustrate the complex regulation of the murine *iNOS* gene and indicate that transcription factors that interact with the *iNOS* promoter do so cooperatively (532). In fact, murine Mϕ can be induced to produce NO even at concentrations of IFN-γ and LPS that -by themselves- are ineffective; in other words, these two mediators appear to function synergistically (536). This seems to suggest that host and pathogen products may induce complementary signals that are required for optimal expression of iNOS and NO generation. This is supported the demonstration that cell treatment with IFN-γ in combination with other cytokines (*e.g.,* TNF-α, IL-1) (537-539); microbial products (*e.g., E. coli* LPS, *M. tuberculosis* mannose-capped lipoarabinomannan, *P. falciparum* GPI) (66, 511, 540); or inert particles (*e.g.,* monosodium urate crystals [MSU]) (541) results in enhanced iNOS expression and NO production, through a mechanism that involves multiple signaling pathways (*e.g.,* Jak2; JNK and ERK1/2 MAPKs; PKC) and transcription factors (*e.g.,* IRF-1, NF-κB, STAT1α). Further study on the identification and interaction of the participating kinases and transcription factors will be essential to fully understand transcriptional regulation of the *iNOS* gene during pathological conditions.

Post-transcriptional and Post-translational Control

In addition to the complex transcriptional regulation of the *iNOS* gene, iNOS expression is also controlled *via* post-transcriptional and post-translational mechanisms. At the post-transcriptional level, it is well known that the amounts of mRNA depend on the equilibrium between the rhythm of their synthesis and their degradation. *In vitro* studies with murine Mϕ demonstrated that TGF-β destabilized iNOS mRNA and accelerated its degradation (542). In human hepatocytes, rat astrocytes and murine Mϕ, the 3'-untranslated region (UTR) of iNOS mRNA was found to possess several adenylate/urydilate-rich elements (AREs) known to destabilize mRNAs, which in turn diminishes the efficacy of the translation process (499, 543, 544). Protein stability is also an important mechanism controlling iNOS levels. Enhanced degradation of iNOS protein by TGF-β and the subsequent suppression of NO production in Mϕ was the first known example of post-translational regulation of iNOS (438). Degradation of the iNOS protein has been shown to involve the proteasome pathway (545). In Mϕ, addition of the proteasome inhibitor lactacystin after induction of the *iNOS* gene by LPS drastically increased the amount of steady-state iNOS protein (546).

As is the case for cNOS, iNOS is active only as a homodimer. The dimerization requires binding of calmodulin and incorporation of heme and possibly Zn^{2+}. For iNOS and nNOS, the dimers are further stabilized by binding of BH_4, one of the cofactors of all NOS, and of the substrate L-arginine

(451). Several proteins block the dimerization and activity of NOS isoforms, including the Mϕ product NAP110 (which has 70% amino acid homology to a tumor cell protein that inhibits iNOS) and the central nervous system protein kalirin, which also inhibits iNOS and might protect the nervous tissue during inflammatory processes (547, 548). Cytokines such as IFN-γ, TNF-α, IL-1, IL-4, and TGF-β induce or suppress guanosine triphosphate cyclohydrolase I, the key enzyme of BH_4 synthesis, which is essential for NOS catalysis (549).

All NOS isoforms can be phosphorylated within cells (550). Serine phosphorylation of eNOS by the Akt kinase was shown to be a prerequisite for activity (551). In Mϕ, treatment with PTP inhibitors increased the levels of tyrosine phosphorylation of the iNOS protein (552). Moreover, Mϕ pretreatment with PI-3-K and p70-S6 kinase inhibitors reduced LPS-mediated iNOS phosphorylation and NO production (553). These studies suggest that the phosphorylation status of iNOS plays an important role on iNOS activity and thus, in NO generation.

Another factor that determines NOS activity is the availability of its substrate, L-arginine. Mϕ and VSMC can regenerate arginine from citrulline for NO production. Argininosuccinate synthetase, the rate-limiting enzyme of the citrulline-NO cycle can be induced by LPS ± IFN-γ in the same cells as iNOS (554, 555). High-output NO production depends on extracellular L-arginine even when an adequate level of intracellular arginine is present. In most cell types, uptake of L-arginine is mediated by a family of cationic amino acid transporter proteins (CAT)1, CAT2A, CAT2B and CAT3. In Mϕ, CAT1 and CAT2A were up-regulated by LPS stimulation, and CAT2-/- mice showed more than 90% reduction of arginine uptake and NO production in response to IFN-γ + LPS (556). Extracellular arginine concentration is strongly modulated by arginase. This enzyme degrades arginine to urea and ornithine and exists in at least two isoforms, arginase I and II (557). In Mϕ and bone marrow-derived DCs, TGF-β and Th2 cytokines (*e.g.*, IL-4 ± IL-10) have been found to strongly increase arginase I or arginase II (558). Up-regulation of arginase before Mϕ iNOS induction by IFN-γ + TNF-α or LPS prevented NO production by substrate depletion (559). These data indicate that arginine transport and availability along with iNOS activity form a functional (and perhaps structural) unit (442).

CHAPTER 4
THE CHEMOKINE SYSTEM

Chemokine Structure and Classification

The term "chemokines", an abbreviation of "chemoattractant cytokines", was created to describe a family of related proteins that are effective at inducing leukocyte chemotaxis *in vitro* and recruitment *in vivo*. However, it is now clear that chemokines have a broader range of activities, including activation of various leukocyte functions (560). Chemokines are highly basic proteins of 70-125 amino acids with molecular masses ranging from 6 to 14 kDa. In most situations, chemokines are thought to act as monomers and are usually secreted, although some are expressed on the cell surface (561). Even though their sequence identity can be quite variable (25-95% identity), their overall tertiary structure is strikingly similar (562). Most chemokines contain at least four cysteines that form two disulfide bonds, one between the first and the third and one between the second and the fourth cysteine. The resulting structure contains three β-sheets with short loops in a Greek key formation (561).

Chemokines can be classified by structure according to the number and spacing of conserved cysteines (C) into four major groups, given the preferred names CXC, C, CC, and CX_3C, which are used in the systematic nomenclatures. Less commonly these groups are referred to by the Greek letters α, β, γ and δ, respectively. CXC, CC, and CX_3C chemokines have four conserved cysteines, whereas C chemokines have only two, corresponding to the second and fourth cysteines in the other groups. A small subgroup of CC chemokines has six cysteines. Whereas the first two cysteines of CC chemokines are adjacent, CXC and CX_3C chemokines are distinguished by the presence of one (CXC) or three (CX_3C) amino acids between the first and the second cysteines, (563). CXC chemokines can be further subclassified by structure into ERL+ and ERL- molecules based on the presence or absence of the tripeptide motif glutamic acid-leucine-arginine (ERL) N-terminal to the first cysteine. ERL+ CXC chemokines are specific for myeloid cells, whereas ERL- chemokines attract a variety of leukocyte subsets (561). The two regions of each chemokine that interact with the receptor are an exposed loop in the backbone between the second and third cysteine -believed to be required for low affinity binding-, and the N-terminal portion before the first cysteine, which represents the region of most variability. The N-terminal binding site is required for receptor signaling upon ligation, and the length and amino acid composition of the N-terminus determines whether a chemokine binds with high affinity to a receptor and whether binding has agonistic versus antagonistic effects (562). Systematic chemokine names are built from cysteine subclass roots, followed by "L" for "ligand" and a number (*e.g.*, CCL1, CXCL8), according to the new classification system of chemokines proposed by Zlotnik and Yoshie in 2000 (564).

CHEMOKINE RECEPTORS

Structure and Classification of Chemokine Receptors

The specific effects of chemokines on their target cells are mediated by members of a family of membrane-bound receptors coupled to G-proteins and composed of 7-transmembrane domains (565). Chemokine receptors are part of a much bigger superfamily of G-protein-coupled receptors (GPCR)s that include receptors of hormones and neurotransmitters. Major hallmarks of chemokine receptors are as follows: i) a length of 340 to 370 amino acids; ii) a short extracellular N-terminus is acidic overall and may be sulfated on tyrosine residues and contain N-linked glycosylation sites; iii) 7 α-helical transmembrane domains -with 3 intracellular and 3 extracellular connecting loops composed of hydrophilic amino acids- are oriented perpendicularly to the plasma membrane; iv) a cysteine in each of the 4 extracellular domains; v) a disulfide bond links highly conserved cysteines in extracellular loops 1 and 2; vi) an intracellular C-terminus contains serine and threonine residues that act as phosphorylation sites for receptor regulation; vii) G-proteins are coupled through the C-terminus segment and possibly through the third intracellular loop. Although most chemokine receptors recognize more than one chemokine, they are almost restricted to a single subclass. Thus, the nomenclature system is rooted by the chemokine subclass specificity of the receptor. CC and CXC chemokine receptor names consist of the root CCR and CXCR, respectively, followed by a number (563, 566).

Regulation of Chemokine Receptor Expression

Chemokine receptors are defined as either homeostatic or inflammatory, and this classification is essentially based on function rather than function and expression (567). Most of the chemokine receptor genes are constitutively expressed in cells and cell surface expression varies according to the receptor ranging from ~1000/cell (*e.g.*, CXCR2) to ~40000/cell (*e.g.*, CXCR4) (568, 569). There are two major mechanisms for chemokine receptor regulation on the cell surface: desensitization and altered gene expression. As with other GPCR, chemokine receptors undergo transient homologous (*via* direct binding) and heterologous (*via* transactivation) desensitization leading their internalization (570, 571). This process is thought to involve phosphorylation of serine and threonine residues in the C-tail of the receptor by GPCR kinases and receptor sequestration by internalization. Desensitization might be critical for maintaining the capacity of the cell to sense a chemoattractant gradient (566). The second mechanism -altered gene expression- is particularly evident in the differential trafficking of T cells to lymph nodes. Naïve T cells express high levels of homeostatic receptors that mediate recirculation through the secondary lymphoid organs. Upon clonal selection, the homeostatic receptors are down-regulated and inflammatory chemokine receptors are up-regulated on the effector cells (572). This allows the effector cells to migrate into tissues where the ligands for the inflammatory receptors are expressed (567).

Chemokine receptors fall into two categories according to their expression patterns, those expressed exclusively on a small number of leukocyte types and those that are more broadly expressed (561). Moreover, some receptors are present on most leukocytes but appear to be functionally significant only for specific subpopulations. For instance, both CXCR1 and CXCR2 are expressed on most leukocytes but are functional only for Nϕ, monocytes/Mϕ and mast cells (563). Chemokine receptor expression is regulated both positively and negatively by a variety of pro- and anti-inflammatory stimuli, including IL-2, IFN-γ, the CD3 complex, IL-4, IL-10, and TGF-β (561, 573-575) as well as in response to pathogen invasion, including *Leishmania* (576), HIV-1 (577), Influenza virus (578), and *Chlamydia* (579). In addition, their expression can also be modulated by cell activation and/or maturation. Only activated effector lymphocytes usually respond to inflammatory chemokines because naïve cells typically do not express receptors for these chemokines (575). For example, T cell activation via TCR stimulation triggers increased expression of CCR7, CCR8 and CXCR5 (580), whereas monocyte-derived DC maturation switches off expression of receptors for inflammatory chemokines and increases expression of lymph node-homing chemokines (581). Thus, expression of chemokine receptors is an axis of regulation that can greatly influence leukocyte trafficking (561).

Signaling through Chemokine Receptors

The functional consequences of chemokine receptor activation are not limited to cell locomotion (*e.g.*, cytoskeleton reorganization, cell polarization), but they also affect gene transcription, cell proliferation, degranulation and oxidative burst. The initial events in chemokine-induced signal transduction determine the outcome of the response and must take place in the proximity of the receptor (582). Intracellular signaling by chemokine receptors depends on coupling *Bordetella pertussis* toxin-sensitive heterotrimeric G-proteins. During ligand binding, chemokine receptors associate with G-proteins, which then become active and dissociate into Gα and Gβγ subunits. Phospholipase (PL)C activation by Gβγ results in the production of two second messengers i) inositol triphosphate ($InsP_3$), which mediates calcium release from intracellular compartments and ii) diacyl-glycerol (DAG) that acts in conjunction with calcium to activate various isoforms of PKC (566). The activation of PKC and calcium-sensitive protein kinases catalyze protein phosphorylation, which activates a series of coordinated signaling events that eventually lead to cellular responses (583, 584). Another well established effector of $G\beta\gamma$ subunits is $PI_3K\gamma$ (585). Through this enzyme chemokines activate PKB (c-Akt) (586) and ERK1/2 MAPK (587). There is increasing evidence that chemokine receptors can also activate several intracellular effectors downstream of $G\alpha$ coupling, including Ras and Rho proteins (588), PLA_2, PI_3K (589), tyrosine kinases (*e.g.*, Src, Pyk) (590), and the MAPK pathway (591), probably *via* Src (582). Even though most work pertaining to signaling events by chemokine receptors has focused on $N\phi$ and interactions between CXCR1 and CXCR2 with their ligands (561), no doubt the precise signaling mechanisms by which all chemokine receptors regulate cellular function will soon become evident (566).

BIOLOGICAL ACTIONS OF THE CHEMOKINE SYSTEM

The basis for the diversified activities of chemokines rests upon the ubiquitous nature of chemokine production and chemokine receptor expression. Indeed, virtually every cell type can produce chemokines and expresses a unique combination of chemokine receptors. Moreover, chemokine receptors make use of a flexible and complex network of intracellular signaling machineries that can regulate a variety of cellular functions (582). The amazing redundancy in the chemokine system, where receptors bind many different chemokines and receptors are expressed on the same cell type, allows for fine regulation of the immune response. In addition, the selective expression of specific chemokines in distinct tissues by different stimuli results in selective leukocyte infiltrates. The biological activities of chemokines range from the control of leukocyte trafficking in basal and inflammatory conditions to regulation of Th1/Th2 development, hematopoiesis, lymphoid organ development, angiogenesis and wound healing (592). Two of the main functions of chemokines and their receptors are described below.

Leukocyte Migration

Chemokines and chemokine receptors are involved in the two distinct steps of leukocyte recruitment: extravasation and directional migration. Extravasation from the blood into lymph nodes, Peyer's patches, and inflamed tissues is a step determined by several sequentially acting receptor-ligand pairs (593). In an initial process mediated by selectins, leukocytes interact transiently with the vascular endothelium while searching for activating signals from chemokines. The selectins mediate low-affinity interactions, which, in combination with the shearing stress of blood flow, produce the process of "rolling adhesion" (594). Upon binding of the chemokine to its receptor on the leukocyte, integrins are up-regulated leading to firm adhesion to the vessel wall and transmigration (595). Thus, while chemokine/chemokine receptor interaction represents an essential component, the specificity of leukocyte extravasation rests on a combinatorial effect resulting from the expression of three determinants: selectins, chemokine receptors and integrins (594).

The second step controlled by chemokines and their receptors is the directional migration and positioning of leukocytes within secondary lymphoid organs and tissues. As the cell moves forward the affinity of the adhesion molecules at the rear of the cell decreases. The migration process continues along a concentration gradient until the leukocyte reaches the source of the chemokine (596). In these conditions, cells may encounter multiple chemoattractant signals in a complex spatial and temporal pattern. For instance, it has been shown that Nϕ can use chemokine receptors sequentially to move along different chemotactic gradients, a process defined as "multistep navigation" (597). The switch of the use of one type of receptor to another is determined by the capacity of the first to undergo ligand-induced desensitization. Thus, as a general rule, leukocytes express several types of chemokine receptors. These are required to perform distinct steps in the process of extravasation and positioning within the tissue. The expression of specific chemokines, receptors, and adhesion molecules contributes to the selective migration and tissue specificity of the leukocyte (570).

Immune Response

Once considered as separate, parallel systems, the innate and adaptive immune responses are now known to be two interdependent parts of a single integrated system (598). The innate immune response not only constitutes the first line of defense against pathogens, but also provides the "danger signal" that instructs the adaptive immune system to mount a response (599). In turn, the adaptive immunity calls on the innate immune system to provide professional phagocytes (*e.g.*, Mϕ, Nϕ) and specialized granulocytes (*e.g.,* eosinophils, basophils) essential for host defense (598). It is becoming clear that chemokine system has the crucial role of linking innate and adaptive immune responses. Pathogen recognition leads to the release of chemokines from resident tissue cells (*e.g.,* DCs, Mϕ) (600, 601) and alters the expression of their chemokine receptors (576, 602). These events -mediated by TLRs- are essential for both the recruitment of DCs to sites of pathogen entry and to lymph nodes to activate naïve T and B cells through antigen presentation, thereby generating an adaptive immune response. Importantly, chemokines released by resident tissue cells after TLR activation guide these activated T cells into the site of pathogen entry and/or replication. In this way, chemokines link innate immune cell activation in the tissue to the recruitment of antigen-specific T cells generated in the lymph nodes (603). The chemokines and chemokine receptors as well as the sequence of events required for the development of an immune response are summarized in Figure 1 (598).

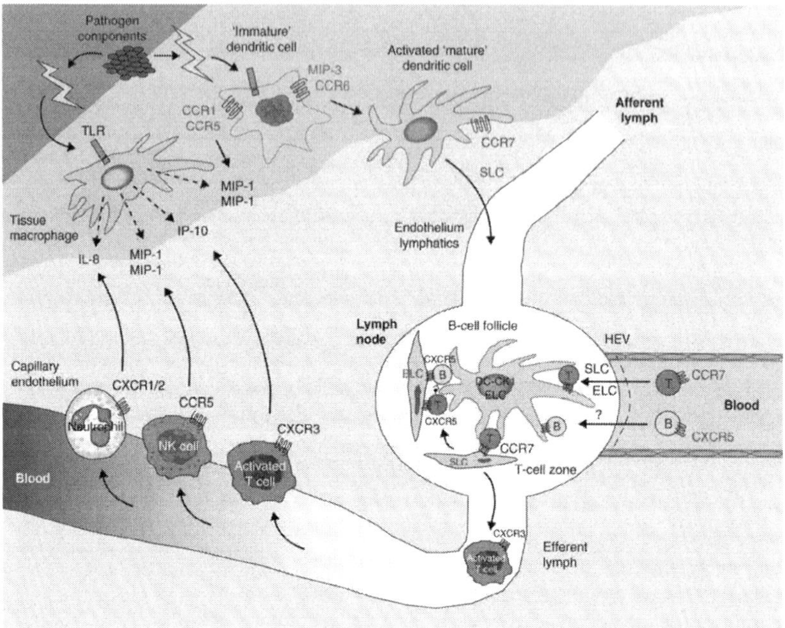

Figure 1. The Chemokine System Orquestrates the Immune Response. Reprinted from *Current Opinion in Immunology*, Vol 14(1):129-135, Luster, A. D., The role of chemokines in linking innate and adaptive immunity, Copyright 2002, with permission from Elsevier. Reference (598).

During the initial events of the immune response, pathogen-induced Mϕ chemokines, including MIP-1α and MIP-1β attract immature DCs from the tissue (604). Upon TLR-mediated DC activation, these cells undergo maturation (605). This is accompanied by CCR5 expression down-modulation and CCR7 up-regulation, which allow DCs to leave the tissues and migrate to lymph nodes (602, 603). The expression of chemokines by lymph-node-stroma and DCs coordinates the juxtaposition of antigen-loaded DCs with the recirculating T and B cells (606). From the activated T cell population, some up-regulate CXCR5, which brings them toward the B cell follicle to instruct B cells in class switching (607). Other T cells that are activated in regional lymph nodes after encounter with activated antigen-loaded DCs are recruited to sites of inflammation by sensing chemokine gradients established at these local sites (603). The early local production of chemokines in the inflamed or infected tissue is essential in shaping the type of immune response. For instance, during CMV infection, early release of MIP-1α and MIP-1β has been shown to be vital for the initial influx of IFN-γ-secreting NK cells in the liver (608), and similar responses have been detected in DCs from *T. gondii*-infected tissue (609). IFN-γ release is then followed by the local production of IFN-γ-inducible chemokines such as IP-10, I-TAC and Mig that attract CXCR3-expressing antigen-specific $CD4^+$ and $CD8^+$ T cells into infected tissue (610). These examples illustrate the requirement of a "chemokine-cytokine-chemokine" cascade for effective protective immunity, which probably represents a general principle that links innate to acquired immunity (598). In addition, during secondary immune responses, chemokines (*e.g.,* MIP-1α, MIP-1β, RANTES) induced by antigen-specific T cells (*e.g.,* cytotoxic $CD8^+$ T cells) recruit innate immune cells (*e.g.,* monocyes, NK cells, granulocytes) into sites of inflammation, serving to amplify the adaptive response with innate immune effector cells (611).

THE CHEMOKINE SYSTEM IN DISEASE

Chemokines and their receptors play critical roles in assuring leukocyte migration to the proper environments for maturation, and to the sites of infection to deal with pathogen invasion. However, high expression can also create the potential for persistent or inappropriately exacerbated response. In fact, recruited activated leukocytes may induce higher expression of chemokines, creating a feed-forward scenario that results in more extensive tissue damage. In this way, the protective effects of leukocytes can be subverted in a manner that causes disease (612). The chemokine system mediates many pathological conditions, including allergies, autoimmune diseases, cancer, infections, inflammatory diseases, transplant rejection and vascular abnormalities (613). In some cases (*e.g.,* eosinophilic pneumonia), one ligand-receptor pair may produce a pathologically straightforward disease (614); however, in most cases, temporal and spatial patterns of chemokine and receptor expression are interwoven with ligand-receptor specificity to result in a multicomponent pathophysiological response (612). To illustrate the multifactorial contribution of chemokines and their receptors to the development of a specific pathology, various examples of their roles in parasitic diseases will be described below.

Parasitic Infections

A diverse range of parasites interacts with the chemokine–receptor network, including the protozoan parasites *Leishmania*, *Toxoplasma*, *Trypanosoma*, *Entamoeba* and *Plasmodium*. Such interactions contribute to determine the parasite-host relationship because of the different roles that chemokines and their receptors can play during parasitic infections, such as direct and indirect microbicidal actions and effects that are detrimental to the host (*e.g.,* increase in parasite survival, formation of tissue lesions) (615).

Chemokine induction exerts a beneficial effect in the host by recruiting and activating cellular function. As illustrated in *Trichomonas vaginalis* and *E. histolytica* infections, CXC chemokine production (IL-8, GROα) seems to contribute to Nϕ activation and subsequent reduction of parasite load by the release of microbicidal enzymes (616, 617). During infection with *T. cruzi*, chemokines appear to be particularly beneficial for the host. *T. cruzi* stimulated the synthesis of MIP-1α, MIP-1β, RANTES and MCP-1 by Mϕ (618), and exogenous administration of these chemokines increased parasite uptake and intracellular parasite destruction through a NO-dependent mechanism (618, 619). These findings indicate that chemokines released by Mϕ after *T. cruzi* infection, participate in the control of parasite replication and might play a role in the acute phase of infection (615). In contrast, during African trypanosomiasis –caused by *T. brucei*- chemokine production seems to be detrimental to the host. This disease is characterized by parasite migration to the central nervous system, causing infiltration of inflammatory cells. In *T. brucei*-infected animals, early chemokine production in the brain (MIP-1α, MIP-2 and RANTES) led to accumulation of inflammatory cells and amplified the formation of lesions (620).

The ability of chemokines to favor T cell responses is evidenced in *T. gondii* infections. *T. gondii* strongly increased the expression of two IFN-γ-inducible chemokines, IP-10 and Mig *in vivo* (621). Moreover, neutralization of IP-10 in *T. gondii*-infected mice inhibited T cell influx into tissues and modified the function of T cells at sites of Th1 inflammation (610), suggesting a role for these chemokines in the polarization toward a Th1–type response to *T. gondii*. *In vitro*, *T. gondii* led to the secretion of GROα, IL-8 and MCP-1 in fibroblasts (622), and *T. gondii* antigens caused MIP-1α and MIP-1β release by Nϕ. These chemokines could be involved in the recruitment of leukocytes such as monocytes, NK cells and DCs during infection, but might also attract Th1 cells (623). Thus, they probably participate in the inflammatory reaction observed during acute invasion. In addition, MCP-1 appears to exert an important antiparasitic effect in *T. gondii* since pretreatment of human monocyte-derived Mϕ with MCP-1 inhibited parasite replication (624), suggesting that this chemokine activates effector functions in Mϕ (615).

In cutaneous forms of leishmaniasis (*e.g., L. major, L. mexicana*) a protective role for MCP-1 has also been described. The concentration of MCP-1 in lesions of patients with self-healing localized cutaneous leishmaniasis (LCL) was high and correlated with a reduction in parasite load. In contrast, MIP-1α expression predominated in lesions of patients suffering from diffuse CL (DCL), while MCP-1 levels were much lower (625). Thus, the predominance of MCP-1 production seems to be beneficial to the host, which might be associated to its ability to activate Mϕ effector functions such as oxidative burst, an effect that has not been attributed to MIP-1α (615). In addition, NK cell-activating chemokines (IP-10, lymphotactin and MCP-1) were expressed in lymph nodes early after infection of resistant mice (626), supporting a role for NK cell activation in early parasite control. The importance of Th1 responses during *L. major* infection was supported by two separate studies in which the Th2 phenotype was associated with increased susceptibility to infection (627, 628). The participation of chemokines in protective immunity is also illustrated during visceral leishmaniasis, which is caused by *L. donovani*. In response to infection, a rapid increase was observed in hepatic production of MIP-1α, MIP-1β and IP-10. Kupffer cells appeared to be important chemokine sources early after infection, whereas T cells seemed to be necessary for the maintenance of maximal tissue chemokine levels, thereby favoring hepatic granuloma formation and acquisition of resistance (629).

Comparison of the proinflammatory capacities of *L. major* and *L. donovani* using an air-pouch model revealed that even though both species led to a similar pattern of chemokine (*e.g.,* MIP-1α, MIP-1β, MIP-1, IP-10, MCP-1) and chemokine receptor (*e.g.,* CCR1, CCR2, CCR5) expression in infected mice, *L. major* was a more potent chemokine receptor inducer. These data suggest that variations in chemokine receptor expression might influence the intensity of the inflammatory response and ultimately contribute to the development of different pathologies (576). In line with this reasoning, cutaneous leishmaniasis is characterized by skin lesions due to a chronic inflammatory reaction that usually restrains the parasite to the site of infection. In contrast, visceral leishmaniasis only causes redness at inoculation sites before the parasite migrates toward the liver and the spleen (630).

The induction and roles of chemokines and their receptors during *Plasmodium* infection have been discussed in Chapter 1. However, it should be noted that the human species *P. vivax* and the simian species *P. knowlesi* gain entry to the RBC by binding specifically to the Duffy antigen receptor for chemokines (DARC). Individuals from equatorial and southern Africa selectively lack Duffy expression on RBC due to a mutation in the promoter region and are thus resistant to *P. vivax* infection (631). DARC binds IL-8, MCP-1 and RANTES, but not MIP-1α and MIP-1β (632, 633). Although the exact physiological role of DARC remains elusive, it has been proposed to function as a chemokine clearance factor to maintain chemokine gradient from tissue blood and as a chemokine-presenting molecule on the surface of endothelial cells (634). The fact that DARC is also a receptor for *P. vivax* suggests that it could be an important therapeutic target (633).

Therapeutic Applications

As illustrated by the numerous pathologies related to chemokine and chemokine receptor expression, selective blockage of one or more components of the system has emerged as a promising therapeutic application. Despite the apparent redundancy of the chemokine-receptor network, it is commonly believed that the system has built several levels of control that lead to specificity *in vivo*; however, the underlying mechanisms are still far from being well understood (635). Several strategies to target chemokine and their receptors are currently under investigation and some have progressed to clinical trials (*e.g.,* MS, HIV, RA, asthma) (636). These approaches include receptor antagonists (*e.g.,* truncated chemokines, small molecule inhibitors) (637), anti-chemokine Abs (638), anti-receptor Abs (639), and DNA vaccines (640). The search for new therapeutic targets using any approach depends on the adequacy of animal models to mirror human disease; however, it is promising that many studies targeting the chemokine receptors have shown that inflammation can be reduced (reviewed by Proudfoot (635)). Moreover, if the hypothesis that chemokine system operates in a temporal and spatial sequence of events is correct, it is conceivable that interfering with an essential link in the chain might result in total inhibition of the inflammatory process. A better understanding of the mechanisms that control changes on chemokine and chemokine receptor expression in the various leukocyte types that participate in the development of a given disease will contribute to identify key steps that could be targeted therapeutically (635).

CHEMOKINE INDUCTION

Expression of inducible chemokines can be elicited by almost any stimulus that alters cellular homeostasis and mRNA encoding induced chemokines can increase over 300-fold within a few hours of activation. Thus, inducible chemokines can be thought of as a cellular "SOS response" that recruits leukocytes to areas of tissue injury (612).

In Response to Cytokines

Chemokines are typically expressed in one of two characteristic patterns. Those involved in homeostatic trafficking are expressed constitutively by many cell types in tissue-specific sites and contribute to homeostatic homing in these areas. In contrast, the expression of inflammatory chemokines is induced only under specific conditions, typically by inflammatory cytokines (561). In general, IL-1β and TNF-α induce broad expression of inflammatory chemokines by a variety of cell types (*e.g.,* epithelial, endothelial, stromal cells, leukocytes). Classical examples include eotaxin, IL-8, IP-10, MCP-1, MIP-1β and RANTES. These chemokines are produced in different types of inflammatory diseases, possibly with distinct kinetics (Reviewed in (594)). Th1-like responses produce inflammatory reactions characterized by interactions between Mϕ, Nϕ, and Th1 cells that produce IL-12 and IFN-γ. These cytokines induce the expression of a number of chemokines (*e.g.,* IL-8, IP-10, Mig) that act to further recruit monocytes, Nϕ and Th1 cells (561). Th2-like responses produce allergic reactions involving basophils, eosinophils, mast cells and Th2 cells that produce IL-4 and IL-13 (641). These cytokines up-regulate –among others- eotaxin, MCP-1, MDC, TARC and TCA-3 production, which leads to a Th2 pattern of cell recruitment. Moreover, cytokines such as IFN-γ and IL-4 can antagonize each other's chemokine induction (642), thereby favoring a specific type of immune response.

In Response to Pathogenic Components

In addition to cytokine-mediated chemokine production, their induction is also regulated by numerous pathogens. It is now widely recognized that chemokine induction occurs directly in response to pathogenic components via TLR activation (643). Professional phagocytes such as Mϕ express TLRs, which function as pattern-recognition receptors (PPRs) that recognize conserved molecular patterns shared by a large group of microorganisms, and are essential for translating the recognition of microbial components to activation of the immune system (644). TLR ligation triggers intracellular signaling, which ultimately ends up in the induction of a subset of chemokines (Figure 2) (598). Of interest, comparative studies showed some differences in the patterns of chemokine induction depending on the TLR that is activated (601, 645). For instance, in DCs, MIP-1α, MIP-1β and RANTES were found to be induced by agonists of both TLR2 and TLR4, whereas IP-10 was preferentially induced by TLR4 agonists (*e.g., E. coli* LPS). In contrast, IL-8 was preferentially produced in response to TLR2-specific agonists (*e.g., S. aureus* peptidoglycan and yeast zymosan). These data suggest that pathogens can determine the nature of the immune response though differential activation of TLRs and the subsequent patterns of chemokine expression and leukocyte recruitment to the sites of infection and/or inflammation (598). Future investigation will contribute to better define the role of selective TLR activation in the development of a specific pathology. Table 1 illustrates some examples of pathogen-inducible chemokine expression through TLR-mediated intracellular signaling.

Table 1. Chemokine Induction *via* TLR Activation

PATHOGEN/COMPONENT	CHEMOKINE(S)	TLR(S)	SIGNALING EVENTS	CELL TYPE /ORGAN	REF.
Bacteria					
E. coli DNA	IL-8	TLR4	ERK1/2 MAPK	epithelial cells	(646)
E. coli LPS	IP-10, MCP-1, MCP-5, MIP-α, MIP-2	TLR4		Mϕ	(645)
	IL-8, IP-10, MIP-1α, MIP-1β, RANTES	TLR4	p38 MAPK, PI3K	DC	(601)
	IL-8	TLR4	p38 MAPK	Nϕ	(647)
E. coli LPS + IFN-γ*	IL-8	TLR4	MD-2, MyD88	monocytes	(648)
	IP-10	TLR4		fibroblasts	(649)
H. pylori	GROα, IL-8, MIP-3α	TLR2, TLR5	NF-κB	epithelial cells	(650)
Neisseria gonorrhoeae Lip lipoprotein	IL-8	TLR2, TLR1	NF-κB	embryonic kidney cells	(651)
Porphyromonas gingivalis	IL-8	TLR2		epithelial cells	(652)
P. gingivalis LPS	IP-10, MIP-1α, MIP-2	TLR2		Mϕ	(645)
Pseudomonas aeruginosa flagellin	IL-8	TLR5	NF-κB	epithelial cells	(653)
Salmonella typhimurium flagellin	GROα, GROβ, GROγ, IL-8, MCP-1, MIP-1α	TLR5		DC	(654)
S. aureus	IL-8	TLR2	NF-κB	keratinocytes	(655)
S. aureus PGN	IL-8, MIP-1α, MIP-1β, RANTES	TLR2	p38 MAPK, PI3K	DC	(601)
S. aureus PGN + IFN-γ*	IL-8	TLR2	p38 MAPK	Nϕ	(647)
	IL-8	TLR2		PBMC	(649)
	IP-10	TLR2		fibroblasts	
Fungi					
Aspergillus fumigatius	MIP-2	TLR2, TLR4	NF-κB	Mϕ	(656)
Candida albicans	MIP-2	TLR2	NF-κB	Mϕ	(657)
Parasites					
L. major LPG	IL-8	TLR2		embryonic kidney cells	(658)
T. cruzi Tc52	IL-8, MCP-1, MIP-1α	TLR2		DC	(659)
IFN-γ + *T. cruzi* tGPI-mucins¶	MCP-1	TLR2		pleural cavity	(660)
Virus					
HSV-1	MCP-1	TLR2		brain	(661)
RSV + LPS*	IL-8	TLR4		epithelial cells	(662)
Other Stimuli					
CpG-rich oligonucleottides	IL-8	TLR9		epithelial cells	(662)
DsRNA	IP-10	TLR3		PBMC, fibroblasts	(649)
Fibrinogen	MCP-1, MIP-1α, MIP-2	TLR4		Mϕ	(663)
Nϕ elastase	IL-8	TLR4		embryonic kidney cells	(664)

*Synergistic effect,

¶Potentiate cellular responses to IFN-γ

dsRNA, double-stranded RNA; HSV, herpes simplex virus; LPG, lipophosphoglycan; PGN, peptidoglycan; RSV, respiratory syncytial virus.

Mechanisms of Regulation of Chemokine Expression

Transcriptional Control

Chemokine expression is primarily regulated at the transcriptional level (665). As summarized in Table 2, the involvement of multiple kinases and transcription factors has been reported in chemokine induction by a variety of stimuli (*e.g.*, cytokines, pathogens, ROS) (see also Table 1). In the promoter regions of numerous chemokine genes, functional binding sites for diverse transcription factors have been identified. To cite some examples: NF-κB binding sites were reported in the murine chemokine genes of *MCP-1* (666), *MIP-1α* (667) and *MIP-2* (668) as well as of human *IL-8* (669) and *MCP-1* (670); AP-1 was found to bind to the human *MCP-1* (671) and *IL-8* (672) genes; functional binding sites for cAMP-response element-binding protein (CREB) were identified in the murine *MIP-1β* (673) and human *RANTES* (674) promoters; C/EBP has been shown to bind the promoter regions of *IL-8* (669) and murine *MIP-1α* (667); and potential STAT1α binding sites were reported in the human (675) and murine (676) *IP-10* genes. Importantly, the cooperative binding of various regulatory elements forming an enhanceosome (transcription factors, coactivators) appears to be required for maximal activation of some chemokine promoters (*e.g.*, *MCP-1, RANTES, IL-8*) (669, 670, 677, 678).

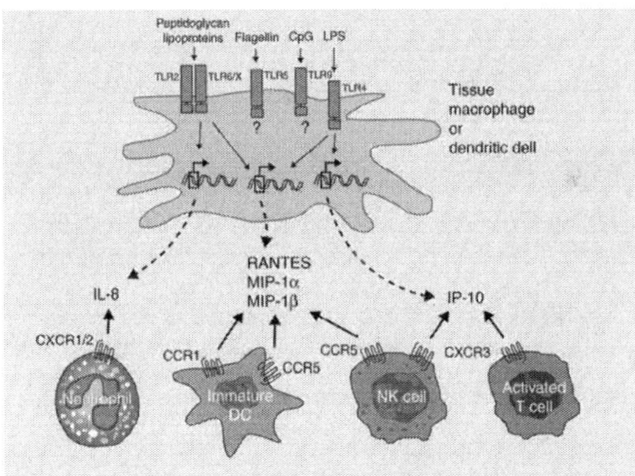

Figure 2. Chemokine Induction *via* TLR Activation. Pathogens selectively activate distinct TLRs expressed on resident tissue Mϕ and DC. For instance, TLR4 is activated by LPS from Gram-negative bacteria. Activation of different TLRs induces the expression of distinct sets of chemokines. Activation of TLR2 induces the expression of IL-8, RANTES, MIP-1α and MIP-1β, whereas activation of TLR4 induces the expression of IP-10, RANTES, MIP-1α and MIP-1β. In turn, these chemokines recruit different subsets of leukocytes. Reprinted from *Current Opinion in Immunology*, Vol 14(1):129-135, Luster, A. D., The role of chemokines in linking innate and adaptive immunity, Copyright 2002, with permission from Elsevier. Reference (598).

Table 2. Signaling Events Involved in Chemokine Gene Expression

STIMULUS	CHEMOKINE	SIGNALING EVENTS	CELL TYPE	REF.
Cytokines				
IFN-γ	MCP-1	STAT-1α, Sp1*	astrocytoma cells	(679)
	Mig	NF-κB	Mϕ	(680)
IL-1β	IL-8	PI3K-mediated AP-1 and NF-κB; p38-mediated AP-1	glial cells	(681)
	IL-8, MCP-1	p38 MAPK	epithelial cells	(682)
	MCP-1	AP-1, NF-κB*	endothelial cells	(670)
IL-4	Eotaxin	STAT6	hepatocytes, endothelial cells	(683)
TNF-α	IL-8	Ras- and MEKK-1-dependent ERK1/2, JNK, AP-1 and NF-κB	epithelial cells	(684)
	MCP-1	Rac-1/NOX/ROS/NF-κB	endothelial cells	(685)
	MCP-1	ERK1/2 MAPK- and ROS-dependent	VSMC	(686)
Pathogens				
E. coli LPS	MIP-2	c-Jun, NF-κB*	Mϕ	(687)
Haemophilus influenza	IL-8	Src/Raf-1/MEK1/2/ERK1/2; p38 MAPK	epithelial cells	(688)
HCV + HIV proteins*	IL-8	p38 MAPK, SHP2 PTP, AP-1	hepatocytes	(689)
H. pylori	MCP-1, IL-8	Ras, ERK1/2, JNK and p 38 MAPK, NF-κB	epithelial cells	(690)
HIV-1 Tat protein	IL-8, MCP-1	EKR1/2 MAPK	astrocytes	(691)
	IP-10	p 38 MAPK		
HSV	RANTES	IRF-3, NF-κB	fibroblasts, Mϕ	(692)
Influenza virus	RANTES	JNK and p38 MAPK	epithelial cells	(693)
Leukemia virus Tax protein	IL-8	AP-1, NF-κB*	T cells	(694)
Mycobacterium bovis BCG	IL-8	PTK/NF-κB	monocytes	(695)
M. tuberculosis	MCP-1	ERK1/2 MAPK	monocytes	(696)
RSV	RANTES	AP-1, C/EBP, CREB, IRF, NF-κB*	epithelial cells	(677)
T. gondii	IL-8	IKKβ/IκBα/NF-κB	epithelial cells	(697)
	MCP-1	NF-κB	fibroblasts	(698)
Trichomonas vaginalis	IL-8	ERK1/2 and p 38 MAPK, NF-κB	Nϕ	(699)
Other Stimuli				
Angoitensin II	MCP-1	p 38 MAPK, NF-κB	fibroblasts	(700)
CpG-rich oligonucleottides	MIP-2	c-Jun, NF-κB	Mϕ	(701)
dsRNA	IL-8	PKR, p38 MAPK	epithelial cells	(702)
	RANTES	PKR		
H_2O_2	MCP-1	AP-1	endothelial cells	(671)
	IL-8	AP-1, NF-κB	epithelial cells	(672)
MSU crysrals	IL-8	Src/ERK1/2 MAPK, AP-1, Src/IKK/NF-κB	monocytes	(703, 704)
Uric Acid	MCP-1	ERK1/2 and p38 MAPK, ROS	VSMC	(705)
VEGF	MCP-1	AP-1	endothelial cells	(706)

*Cooperative action

BCG, bacillus Calmette-Guerin; HCV, hepatitis C virus; IKK, IκB kinase; MEKK, MAPK/ERK kinase kinase; MSU, monosodium urate; PKR, dsRNA-dependent protein kinase; TRAF, TNFR-associated factor; VEGF, vascular endothelial growth factor.

As noted in Tables 1 and 2, the expression of many chemokine genes appears to be under the control of MAPK cascades and requires the participation of the transcription factors NF-κB and AP-1. MAPK are a family of serine/threonine kinases that constitute important mediators of signal transduction from the cell surface to the nucleus. Most family members are activated by dual phosphorylation on threonine and tyrosine residues in response to a wide array of extracellular stimuli (*e.g.*, cytokines, growth factors, pathogens, stress), and their regulation has been implicated in diverse physiological processes, including cell proliferation, differentiation and activation (707). MAPKs constitute more than a dozen of proteins belonging to three families, ERKs, p38 MAPKs, and JNK/stress-activated protein kinases (SAPKs) (708-710). All MAPK cascades are organized in a conserved three-kinase architecture consisting of a MAP kinase (MAPK), a MAP kinase kinase (MAP kinase activator, MEK), and a MAP kinase kinase kinase (MEKK), which activates the MAP kinase activator (711).

MAPKs act in concert among them and with other signaling cascades. Therefore, crosstalk between pathways is crucial to the coordinated cell responses. MAPKs may cooperate or antagonize with each other and the integration signals may take place at many levels; that is upstream of the cascades, within the cascades and within the substrates (707). For instance, cAMP either inhibits or synergizes with other stimuli to activate ERKs, depending on the cell type, and probably acts upstream of the kinase cascade (712). Because several protein kinases lie downstream of MAPKs, they can be regulated by more than one family member. This is the case of MAPK-activated protein kinase 3 (MAPKAP3), which may be activated by ERK, JNK/SAPK, or p38, enabling to integrate signals from all three pathways directly (713). Another example is the regulation of the AP-1 transcription factor. Activation of the JNK cascade directly induces phosphorylation of Ser63 and Ser73 on c-Jun (714), which stimulates the transcriptional activity of AP-1 (715), and also prolongs the life of c-Jun by protecting it from ubiquitination by the proteasome (716). ERK1/2 and p38 cascades can also regulate the expression of Jun and Fos constituents of AP-1 complexes (717, 718), and crosstalk with the JNK pathway occurs during activation of ERK1/2 and p38 (719). Thus, the JNK, p38 and ERK1/2 cascades have overlapping and cross-regulatory effects even though they are phosphorylated by distinct MEKs and they differ in their predisposition to activation by specific receptor- and non-receptor-mediated stimuli (717, 719).

Transcription factors are the major targets for MAPKs, which in turn induce transcription of several early genes involved in cell activation, proliferation, and transformation (707). The signaling events involved in MEK1/2-ERK1/2–dependent gene expression include various upstream kinases (*e.g.*, Ras GTP-binding protein, Raf serine/threonine kinase) as well as the transcription factors (*e.g.*, AP-

1, CREB) activated in response to these MAPKs. The MEK1/2-ERK1/2 pathway can directly phosphorylate the transcription factors AP-1 (720), Ets (721) and c-Myc (722), which leads to their activation. Alternatively, MAPKs can phosphorylate and activate downstream proteins named ribosomal S6 kinases (RSKs), which then phosphorylate and activate transcription factors, such as CREB (723).

Instead of a direct phosphorylation, Ras-mediated signals can lead to transcription of NF-κB activation by phosphorylating and activating its positive regulators IκB kinases (IKKs) (724-726). NF-κB is a dimeric transcription factor comprising members of the *Rel* gene family of DNA-binding proteins. NF-κB can modulate the expression of target genes encoding chemokines, growth factors, cytokines and other transcription factors, by binding to the *cis*-acting element contained in their promoter regions (727). Many target genes of NF-κB play important roles in proliferation, prevention of apoptosis, angiogenesis, metastasis, and immune responses (728). Some of the most ubiquitous members of the NF-κB family identified so far include RelA (p65), Rel B, c-Rel, p105/NF-κB1 and p100/NF-κB2 (729). p105/NF-κB1 and p100/NF-κB2 are inactive precursors, which are processed into their active forms of p50 and p52 -respectively- before dimerization. Different subunit combinations (homodimers and heterodimers) have diverse functions in regulating transcription. The heterodimer comprising p50 and Rel A is the most frequently observed NF-κB complex. It is a potent activator of gene expression from κB sites because of the presence of two transactivating domains in the C-terminal region of Rel A (730).

Before stimulation, NF-κB is normally sequestered in the cytoplasm by binding to inhibitory protein κB (IκB)α, IκBβ or IκBε inhibitors (731). This binding masks the nuclear localization signal on NF-κB, thereby preventing its translocation to the nucleus. The inhibitory effects of IκBs on NF-κB activity can be eliminated by the multisubunit IKK cascades (732). The IKK core complex contains four IKK subunits: one IKKα, one IKKβ and two IKKγs. IKKα and IKKβ are the catalytic subunits, both of which can phosphorylate IκBs at specific serine residues in their N-terminal domain. For instance, they can phosphorylate IκBα on Ser32 and Ser36 (733). Phosphorylated IκBs can be recognized by the IκB-E3 ubiquitin ligase SCF (Skp1-Cullin-F-box protein ligase complexes) and degraded through the proteasome pathway (734). IKKγ -also known as NF-κB essential modulator (NEMO)- is required for stabilization of the IKK complex (735). Kinases including NF-κB-inducing kinase (NIK) and Raf-activated MAPK/ERK kinase kinase-1 (MEKK1) phosphorylate subunits of the IKK complex and modulate its activity (725). Even though

NIK and MEKK1 can activate both IKKα and IKKβ, NIK preferentially activates IKKα by phosphorylation on Ser176, while MEKK1 preferentially phosphorylates IKKβ on Ser177 and Ser181 (736). NF-κB transcriptional activity is also regulated through its own phosphorylation by kinases including protein kinase A (PKA) and mitogen- and stress-activated protein kinase 1 (MSK1) (737, 738). IKK is also able to directly phosphorylate p65 on its C-terminal transactivation domain at Ser536 (739). These phosphorylation events regulate NF-κB activity by modulating its DNA binding, transactivation properties and interactions between transcription factors and regulatory proteins (738).

Another transcription factor that can be activated through the Ras/Raf/MEK/ERK pathway is AP-1. This factor is comprised of Jun and Fos family members. The Jun family consists of c-Jun, JunB and Jun D, which can form homodimers or heterodimers with Fos proteins (740). The Fos family consists of c-Fos, Fos B, Fra1 and Fra2, which do not form homodimers but can heterodimerize with Jun. Both Jun and Fos also dimerize with other transcription factors including members of the ATF/CREB family to transactivate gene expression (741). Regulation of c-Fos activity by Raf signaling occurs at the transcriptional level through the activation of transcription factors Elk-1 and CREB. As mentioned above, Ets family members, such as Elk-1, can be phosphorylated by ERK (721), while CREB can be phosphorylated by ERK via RSKs (723). Phosphorylated Elk-1 binds to the serum-response element (SRE) *cis*-acring element on the *c-fos* gene, while phosphorylated CREB binds to the CRE *cis*-acting element to induce *c-fos* transcription (742, 743). Regulation of c-Jun activity by Raf signaling occurs at both the transcriptional and post-transcriptional levels (744-746). c-Fos, whose expression is also positively regulated by Raf-dependent signals, can form heterodimers with c-Jun and bind the AP-1 site on the *c-jun* promoter region inducing its transcription (744). As mentioned above, c-Jun activity is regulated by JNK through phosphorylation at Ser 63 and Ser73 on its N-terminal A1 transactivating domain (714). Interestingly, the Raf signaling pathway also phosphorylates Jun at these residues, and also at Ser243 in the carboxyl DNA-binding domain of c-Jun to regulate positively (745) or negatively (746) AP-1 activity. In addition, ERK1/2 has also been shown to directly phosphorylate c-Jun at these serine residues (720, 746). Altogether, these different lines of evidence indicate that chemokine gene expression is finely regulated through the interplay of diverse signaling pathways leading to the cooperative actions of transcription factors on the chemokine promoters.

Post-transcriptional Control

In addition to transcriptional regulation of chemokine expression, increased levels of chemokine mRNA may be the result of post-transcriptional mechanisms of control leading to enhanced stability of their transcripts. These regulatory events appear to be related to the presence of AREs in the 3' -UTRs of the chemokine mRNAs, which are implicated in mRNA destabilization (747). AREs have been identified in KC, MIP-1α, MCP-1, MIP-2 and IL-8 mRNAs (668, 748-753), and some variants of AREs have also been found in murine MIP-1β transcripts (668). Even though the mechanisms responsible for the increase in chemokine mRNA stability are not yet completely understood, members of the MAPK family -ERK1/2 and p38- have been found to enhance the stability of certain chemokine mRNAs that bear one or more AREs (*e.g.*, MIP-2, IL-8) (754, 755). Moreover, it has been postulated that stabilization of chemokine transcripts might also be enhanced by specific RNA-binding proteins. HuR, an ubiquitously expressed member of the Hu family of RNA-binding proteins, binds to AREs in mRNAs encoding key regulators of inflammatory responses, including c-Fos and TNF-α (756). Upon HuR binding, its targets display increased mRNA stability and translation of the transcripts (757). Of interest, it was recently reported that HuR binds to eotaxin mRNA, which might explain the increased stability of this chemokine transcript in response to cytokine stimulation (758). Although much still needs to be learned about this process, it is currently believed that after binding to its target mRNAs in the nucleus, the HuR-mRNA complex is transported to the cytoplasm where HuR protects the bound mRNA from degradation (759). Another mRNA-binding protein is the redox-sensitive protein adenosine-uridine-binding factor (AUBF). This protein is known to form exceptionally stable complexes with AREs; thus, preventing mRNA degradation (760). Because many stimuli that increase chemokine mRNA expression are able to activate MAPK-dependent signals and/or to induce oxidative stress, it is plausible that in addition to their role in chemokine transcriptional activation, these intracellular events contribute to enhance the half-life of the chemokine transcripts. Table 3 contains some examples of chemokine post-transcriptional regulation in response to a variety of stimuli.

Table 3. Post-transcriptional Regulation of Chemokines

STIMULUS	CHEMOKINE(S)	MECHANISMS	CELL TYPE	REFERENCE(S)
Cytokines				
IL-1	IL-8	TAK1 MAPKKK/ p38 MAPK	epithelial cells	(761)
IL-1β + PDGF*	MCP-1		mesangial cells	(762)
TNF-α	IL-8	ERK1/2 and p38 MAPK	epithelial cells	(754)
TNF-α ± IL-4*	Eotaxin	HuR mRNA stabilizing protein	epithelial cells	(758)
Pathogens				
Adenovirus	IL-8		epithelial cells	(763)
E. coli LPS	KC	p38 MAPK	Mϕ	(764)
	MIP-1α	p38 MAPK	monocytes	(765)
RSV	RANTES		epithelial cells	(766)
S. typhimurium flagelin	IL-8	p38 MAPK	epithelial cells	(767)
Other Stimuli				
H_2O_2	IL-8	p38 MAPK	monocytes	(768)
	MIP-1α, MIP-2		Mϕ	(751, 769)
Hyperoxia	MCP-1		monocytes	(748)
MBP	IL-8		Nϕ	(770)
Nicotin	MIP-1α	ROS	Mϕ	(771)
PDGF	MCP-1		SMC	(750)
Staurosporine	MIP-2	ERK1/2 and p38 MAPK	Nϕ	(755)
Uric Acid	MCP-1		VSMC	(705)
Vanadium	MIP-2		Mϕ	(772)

*Synergistic effect
MBP, eosinophil granule major basic protein; PDGF, platelet-derived growth factor; SMC, smooth muscle cells.

CHAPTER 5
RATIONALE AND AIMS OF THE STUDY

RATIONALE

Malaria is the most important parasitic disease worldwide, ranking among the major health and developmental challenges nowadays (18). With the lack of an effective vaccine and the increase in insecticide and drug resistance, considerable effort has been undertaken to combat this infection. However, this task is extremely challenging due to the complexity of *P. falciparum* life cycle and to our limited knowledge of the host-parasite interactions. Numerous investigations have focused on identifying the inflammatory mediators (*e.g.,* cytokines, chemokines, NO) produced in response to *P. falciparum* and on defining their antiparasitic and pathogenic roles (91, 95, 97, 98, 186). However, the direct contribution of parasite components to the activation of such immunological responses has not been fully unraveled. In addition to a role proposed for plasmodial GPI (60, 66, 195), different lines of evidence suggest that HZ -a parasitic waste product of Hb catabolism- could also be a potential inflammatory agent. During the symptomatic phase of the disease, large amounts of HZ are released into the circulation after schizogeny, and are rapidly phagocytosed by monocytes, Mϕ and Nϕ (384, 386-389, 773). HZ accumulation has been observed in different organs, such as liver, spleen and brain, where its presence correlates with disease severity (393-395). However, a causative relation between HZ and malaria pathology cannot be inferred from these data. The ability of HZ to induce cytokine and chemokine secretion was demonstrated *in vitro* (61, 65, 196, 198) and its thermoregulatory properties were detected *in vivo* (61). Importantly, the production of Abs against HZ was reported among complicated malaria patients, and such Abs inhibited HZ-inducible cytokine production *in vitro* (205). Altogether, these findings support the idea that HZ possesses proinflammatory activity and therefore, it might be a key component in the development of the clinical manifestations of severe malaria (*e.g.,* anemia, CM, PM). Evaluation of the antigenic properties of HZ *in vivo* as well as the identification and characterization of molecular mechanisms underlying its proinflammatory effects *in vitro* are required to further define the role of HZ in malaria immunopathology. A better understanding of the mechanisms through which *P. falciparum* regulates the host immune response might contribute to identify new therapeutic targets to reduce morbidity and mortality related to malaria.

AIMS OF THE STUDY

The current study has been designed to better define the contribution of HZ to the proinflammatory events related to malaria immunopathology, and to identify the intracellular signals underlying the up-regulatory effects of HZ in the Mϕ, one of the major sources of inflammatory mediators in malaria.

Specific Aim 1

To evaluate the ability of HZ to induce Mϕ NO production alone or in combination with IFN-γ and to define the signal transduction events involved in this cellular regulation

NO overproduction has been proposed as key factor in the development of immunosuppression and CM. Therefore, we were interested to determine whether HZ modulates its synthesis in Mϕ. As previously reported in diverse infectious diseases, pathogenic components are able to act in combination with host cytokines to increase iNOS expression and NO production through the cooperative activation of intracellular signals. Knowing that IFN-γ -a potent Mϕ NO inducer- is produced during malaria and that it plays a crucial role in the development of CM, we investigated the ability of HZ to enhance IFN-γ mediated Mϕ NO generation and identified the transductional mechanisms involved. Our data revealed that even though HZ does not induce NO production in Mϕ *per se*, it enhances IFN-γ-mediated iNOS mRNA and protein expression as well as NO synthesis. These events require ERK1/2 MAPK and NF-κB activation by HZ and IFN-γ which exert a synergistic effect on these signaling pathways. This study was published in 2003 in *The Journal of Immunology* (774).

Specific Aim 2

To assess the proinflammatory properties of HZ in vivo to better define its contribution to inflammatory mediator production during malaria

During *P. falciparum* infection, high levels of proinflammatory molecules correlate with disease severity. Previous studies demonstrated that HZ leads to the release of cytokines and chemokines *in vitro*. The use of the experimental murine air pouch model as well as the intravenous injection of HZ allowed us to evaluate the inflammatory properties of this parasite metabolite *in vivo*, including its ability to induce leukocyte recruitment as well as chemokine, chemokine receptor, and cytokine production. As these immunological responses are implicated in malaria pathophysiology, HZ might emerge as an attractive target for antimalarial therapy. As we reported in 2004 (775), HZ induces leukocyte recruitment *in vivo*, especially of Nϕ and monocytes. In line with this observation, HZ injection increases mRNA and protein levels of various inflammatory molecules including chemokines (MIP-1α, MIP-1β, MIP-2, MCP-1), chemokine receptors (CCR1, CCR2, CCR5, CXCR2, CXCR4), cytokines (IL-1β, IL-6) and myeloid-related proteins (S100A8, S100A9, S100A8/A9). Altogether, these data demonstrate that HZ is a potent proinflammatory agent *in vivo*, which could contribute to the immunopathology related to malaria.

Specific Aim 3

To determine the molecular mechanisms responsible for Mϕ chemokine up-regulation in response to HZ

Even though chemokine production has been reported in HZ-treated monocytes/Mϕ, the mechanisms responsible for this cellular regulation remained unexplored. In this specific aim, we wished to investigate whether HZ controls Mϕ chemokine expression at the transcriptional level and to identify the signaling events involved. In addition, because HZ leads to ROS generation in Mϕ, we examined their contribution to HZ-mediated chemokine induction. A better understanding of the transductional mechanisms through which HZ modulates the host immune response might be helpful in defining specific therapeutic targets to tame proinflammatory mediator overproduction during malaria infection. As summarized in our most recent publication (776), HZ up-regulates Mϕ chemokine gene expression through a mechanism that requires the activation of the transcription factor NF-κB by two different intracellular signals: i) ERK1/2 MAPK phosphorylation and ii) ROS-dependent phosphotyrosine phosphatase (PTP) down-regulation. These data indicate that HZ triggers ROS-dependent and -independent signals leading to Mϕ activation.

Specific Aim 4

To investigate whether the cooperative action of intracellular signals is required for the expression of H_2O_2-inducible Mϕ chemokine genes

ROS generation has been detected during malaria as well as in HZ-laden Mϕ. Knowing that oxidative stress is a potent Mϕ chemokine inducer, in this final aim of the study, we attempted to more deeply explore how ROS control their expression. The ability of ROS –and more specifically of H_2O_2- to activate several signaling pathways, transcription factors and chemokine genes has been studied separately. We hypothesized that simultaneous induction of various chemokine transcripts results from the cooperative action of different signaling cascades and transcription factors. Such a regulatory mechanism might allow rapid and simultaneous activation of several inflammatory genes in response to pathogenic components, as could be the case of HZ. We found that H_2O_2-inducible Mϕ chemokine expression is a multistep and multicomponent mechanism that requires the participation of various signaling cascades (cAMP, ERK1/2 MAPK), which in turn lead to simultaneous activation of several transcription factors (NF-κB, AP-1, CREB). These data show that crosstalk among intracellular signaling pathways results in the formation of a transcriptional complex responsible for maximal chemokine up-regulation. This study was published in 2002 in *The Journal of Immunology* (777).

CHAPTER 6
RESULTS & DISCUSSION

Because of its enormous impact on public health, malaria is one of the major social and economical burdens of humankind. Currently, most efforts to combat this disease are aimed at immunity against infection by targeting parasite proteins in a multicomponent formulation (antiparasitic vaccine) (34). However, it has long been suggested that immunity against severe malaria is partly antiparasitic and partly antitoxic (toxic effects in response to parasite factors) (138, 778, 779). Even though it is now clear that inflammatory mediator overproduction contributes to malaria pathogenesis (91, 95, 98), the mechanisms underlying such regulation remain largely unexplored. Identification of the parasitic components that participate in the activation of the host immune response and a better understanding of their mode of action is of paramount importance to find new therapeutic antimalarial targets.

Different lines of evidence, including its tissue distribution, rapid phagocytosis and interactions with immune cells (61, 65, 196, 198, 384, 386-389, 393-395, 773), suggest that HZ, a metabolic waste product of the parasite, could be an attractive therapeutic target. Therefore, in the present study, we examined the inflammatory activity of HZ and dissected the regulatory mechanisms involved. We found that HZ has the ability to enhance Mϕ responses to IFN-γ (iNOS expression and NO production) and to activate important Mϕ functions (chemokine and ROS induction). The observed effects of HZ did not result from a generalized cellular activation but were rather dependent on specific intracellular signals regulated in response to HZ. Importantly, the proinflammatory properties of HZ were assessed *in vivo* (*e.g.,* chemokine, chemokine receptor and cytokine production), further supporting the idea that this parasite metabolite participates in the activation of the immunological response during malaria infection. These findings will open the avenue for further studies designed to define the role of HZ in the pathogenesis of malaria.

In malaria, increased NO production has been reported in *P. falciparum*-infected patients (142-144, 780, 781) as well as in experimental models of infection (782, 783). Both protective (143, 490, 784, 785) and detrimental (112, 786, 787) roles have been attributed to this mediator. IFN-γ has been identified as an important Mϕ activator during plasmodial infection (788) and thus, central in the regulation of NO synthesis (490). Therefore, in the first part of our *in vitro* studies, we focused on the ability of HZ to modulate NO generation in Mϕ. Even though HZ itself did not induce Mϕ NO production, it significantly enhanced it in response to IFN-γ. This synergistic effect was found to be dependent on the combined action of HZ and IFN-γ in iNOS expression (Figure 1) (774). Analysis of the molecular events revealed the involvement of complementary intracellular signals and transcription factors. IFN-γ induced the phosphorylation of Jak2 and the subsequent activation of the transcription factor STAT1α, two events that were essential in IFN-γ + HZ-mediated NO production. These observations are not surprising given that the Jak2/STAT1α pathway controls iNOS expression both directly, via STAT1α binding to the iNOS promoter (532); and indirectly, by inducing transcriptional activation of the *IRF 1* gene, a key transcription factor in iNOS regulation (534). Moreover, it is known that depending on the gene, either STAT1α or IRF-1 are essential for enabling IFN-γ to synergize with either host or pathogen molecules (*e.g.,* TNF-α, LPS) (526, 530), which could explain the abrogation of HZ + IFN-γ-inducible NO by specific blockage of the Jak2/STAT1α cascade.

Despite the requirement of Jak2-dependent signals, HZ was found to exert its effect on iNOS expression through alternative pathways. In addition to its ability to activate ERK1/2-mediated binding of NF-κB to the murine *iNOS* promoter, HZ enhanced this event in response to IFN-γ (Figure 2) (774). This is in line with a number of studies that have associated synergistic *iNOS* modulation with an up-regulation of NF-κB by IFN-γ in combination with other host cytokines (*e.g.,* TNF-α) (538), microbial components (*e.g., E. coli* LPS) (789), whole pathogens (*e.g.,* herpes simplex virus type 2) (789), or host-derived inflammatory crystals (*e.g.,* MSU crystals) (541). These data suggest that a dual activating effect on NF-κB might be a common mechanism to enhance NO production during pathological conditions.

In addition to increased NF-κB activity, the synergistic induction of inflammatory genes (*e.g., iNOS, IP-10, ICAM-1*) appears largely dependent on the cooperative action of NF-κB with STAT1α and/or IRF-1 (526-529, 790-793). Many studies have shown that the presence of functional κB sites alongside functional GAS or ISRE sites is sufficient for generation of synergy between IFN-γ and TNF-α/LPS (526, 791, 793). For the synergistic action of STAT1α and NF-κB, the available data seem to indicate that the two transcription factors bind independently to their respective sites on the promoter (526), which might generate a surface with enhanced interaction with the basal transcription machinery; however, the information at present is too limited to draw conclusions (527). By contrast, the mechanism underlying the synergistic action of IRF-1 and NF-κB on various genes (*e.g., iNOS, MHC-I, ICAM-1*) has been extensively studied and there is evidence that these factors interact physically (529, 790, 792), perhaps forming a complex with higher avidity for the recognition sites than the transcription factors individually (527). Thus, the possibility that NF-κB contributes to the HZ + IFN-γ-inducible *iNOS* promoter activation by acting in synergy with STAT1α and/or IRF-1 should not be ruled out.

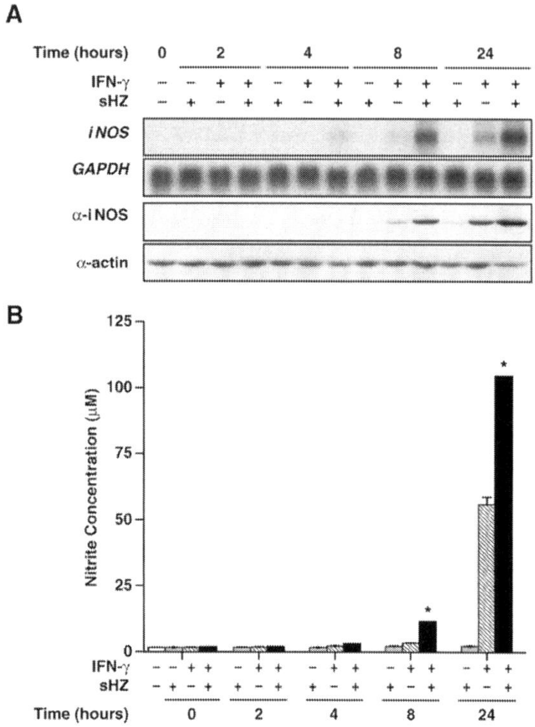

Figure 1. sHZ synergizes with IFN-γ to increase iNOS mRNA and protein levels and to generate NO (time course). (A) B10R Mϕ were stimulated with 25 μg/ml synthetic HZ (sHZ) and/or 100 U/ml IFN-γ for different time periods (0-24 h). Total RNA was extracted and iNOS mRNA levels were monitored by Northern blot analysis (*upper panels*). Following cell stimulation as described above, protein lysates were subjected to Western blotting to evaluate iNOS protein expression (*lower panels*). (B) Nitrite production from either untreated or stimulated (sHZ and/or IFN-γ) cells was measured at various time points (0-24 h) by the Griess reaction. Untreated (open bars); sHZ (gray bars); IFN-γ (stripped bars); sHZ + IFN-γ (solid bars). *, $p < 0.05$, sHZ + IFN-γ vs IFN-γ. Reprinted with permission, from *The Journal of Immunology*. Vol 171:4243-4253. Jaramillo, M., D. C. Gowda, D. Radzioch, and M. Olivier. Hemozoin increases IFN-gamma-inducible macrophage nitric oxide generation through extracellular signal-regulated kinase- and NF-kappa B-dependent pathways. ***Copyright 2003. The American Association of Immunologists, Inc***. Reference (774).

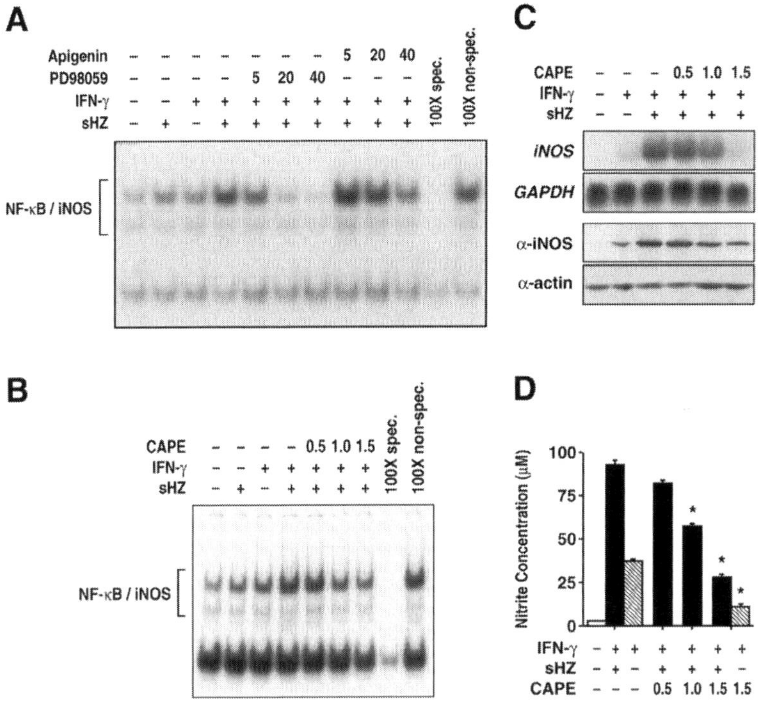

Figure 2. Role of NF-κB on sHZ + IFN-γ-dependent iNOS and NO regulation in Mϕ. Nuclear proteins from sHZ + IFN-γ-stimulated cells pretreated or not with MEK1/2-ERK1/2 inhibitors, apigenin and PD 98059 (μM) (A) or with a NF-kB inhibitor, CAPE (μg/ml) (B) were incubated with a NF-κB/iNOS probe and subjected to EMSA. (C) B10R Mϕ were treated with 0.5-1.5 μg/ml CAPE before stimulation with sHZ + IFN-γ (24 h), and its effect on iNOS mRNA (*upper panels*) and protein (*lower panels*) expression was evaluated by Northern and Western blot, respectively. (D) Following cell stimulation as described in (C), nitrite production was measured by the Griess reaction. Untreated (open bars); sHZ + IFN-γ ± CAPE (solid bars); IFN-γ ± CAPE (stripped bars). * $p < 0.05$, sHZ + IFN-γ + CAPE vs sHZ + IFN-γ; IFN-γ + CAPE vs IFN-γ. Reprinted with permission, from *The Journal of Immunology*. Vol 171:4243-4253. Jaramillo, M., D. C. Gowda, D. Radzioch, and M. Olivier. Hemozoin increases IFN-gamma-inducible macrophage nitric oxide generation through extracellular signal-regulated kinase- and NF-kappa B-dependent pathways. ***Copyright 2003. The American Association of Immunologists, Inc***. Reference (774).

Our results showing the involvement of cooperative signals in the synergistic regulation of iNOS and NO by HZ + IFN-γ are in line with those reported in diverse models of infection. Indeed, immune cell treatment with IFN-γ in combination with bacterial components (*e.g., E. coli* LPS, *M. tuberculosis* mannose-capped lipoarabinomannan) (511, 540) or parasite metabolites (*e.g., P. falciparum* GPI; *T. cruzi* GPI) (66, 217) as well as with whole parasites (*e.g., T. cruzi*) (794) resulted in enhanced NO production, through a mechanism that involves multiple signaling pathways (*e.g.,* Jak2; MAPKs) and transcription factors (*e.g.,* NF-κB, STAT1α, IRF-1). From a biological point of view, synergistic *iNOS* regulation has been proposed as an efficient double-lock system that limits the production of toxic doses of highly reactive molecules to the sites of infection (527). In fact, we and others have observed that whole pathogens (*e.g., T. cruzi*) (794) or host-derived inflammatory particles (MSU crystals) (541) do not activate Mϕ NO synthesis. However, they significantly enhance it in presence of IFN-γ, providing what has been called the second or "triggering" signal for activation (795). In this context, it is conceivable that the requirement for a dual effect on NF-κB -along with the activation of STAT1α- in the synergistic activation of the iNOS promoter by HZ and IFN-γ reflects an example of a more generalized mechanism found in nature, which alerts the immune system against pathogen invasion.

Figure 3 summarizes our current understanding of the signaling events involved in the transcriptional activation of the murine Mϕ *iNOS* gene by HZ + IFN-γ. However, important steps in this regulatory process still remain to be defined, including: (i) the involvement of other transcription factors, such as IRF-1, C/EBPβ and AP-1, which are also controlled by ERK1/2 and play a role in NO regulation (796-799); (ii) the cooperative interactions between NF-κB and other transcription factors, mainly STAT1α and IRF-1; (iii) the role of other second messengers, which - in addition to Jak2 and ERR1/2- also modulate Mϕ NO production (*e.g.,* cAMP) (513); iv) importantly, the Mϕ receptor(s) that interact with HZ remain elusive. Our data indicated that TLR4 was not required for the observed action of HZ on Mϕ NO generation (774). Similarly, our preliminary data revealed that TLR2 was not involved either (data not shown). However, as two separate studies indicated TLR-dependent signaling (*i.e.,* MyD88, TLR9) in response to *Plasmodium* (78, 79), the participation of other members of the TLR family in HZ-mediated Mϕ NO synthesis should not be ruled out; (vi) previous data from our laboratory indicate that up-regulation of Mϕ PTP activity is involved in the attenuation of NO production (800), whereas PTP inhibitors appear to have the opposite effect (452). Because we found that HZ reduces Mϕ PTP activity (776), the participation of this signaling event in HZ-mediated increase in NO synthesis deserves further investigation.

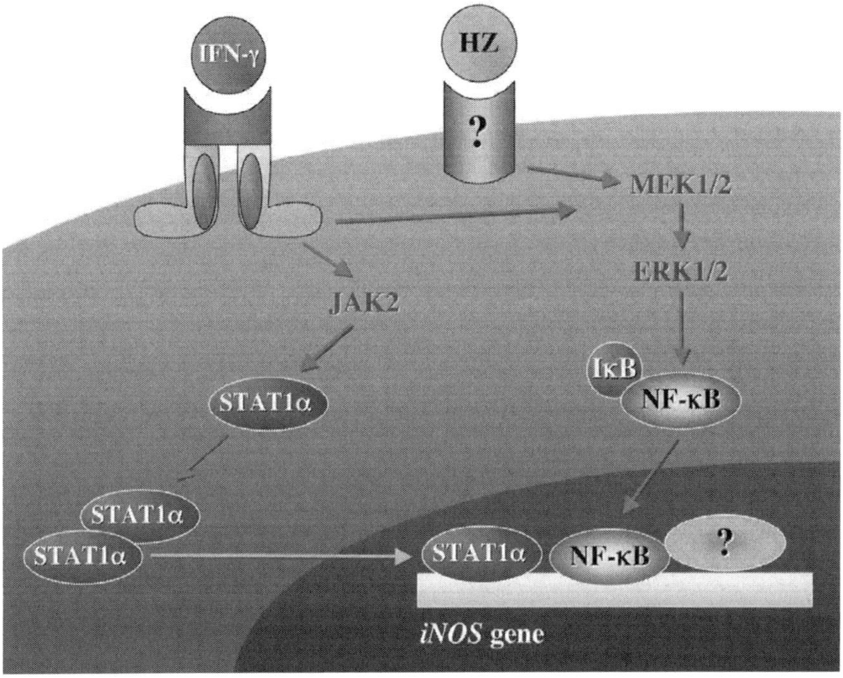

Figure 3. Signaling Events Involved in Hemozoin and IFN-γ-mediated *iNOS* Expression in Macrophages. Reprinted with permission from Landes Bioscience. Olivier, M. and M. Jaramillo. Modulation of positive signaling and proinflammatory responses by hemozoin, a *Plasmodium* metabolic waste. In: Denkers, E. Y. and R. T. Gazzinelli, eds. Protozoans in Macrophages. Austin. Landes Bioscience. 2007: 67-83.

In addition to regulating iNOS at the transcriptional level, it is plausible that through the activation of the ERK1/2 MAPK pathway, HZ + IFN-γ also control iNOS expression *via* post-transcriptional mechanisms. Indeed, there is evidence that ERK1/2 might be involved in the regulation of iNOS translation. PHAS-1 is a protein that binds to translation initiation factor 4E (eIF-4E) and inhibits its function, thus limiting translation (801). ERK1/2 phosphorylates PHAS-1, which inhibits its binding to eIF-4E (802). Thus, the effect of blocking the ERK1/2 pathway might reflect the suppression not only of iNOS transcription but also of iNOS translation in response to HZ + IFN-γ.

Besides protein translation, another factor that determines iNOS activity is the availability of its substrate, L-arginine. High-output NO production depends on extracellular L-arginine even when an adequate level of intracellular L-arginine is present. In most cell types, uptake of L-arginine is mediated by a family of transporters (CAT)s (442). In Mϕ, L-arginine uptake was blocked in presence of ERK1/2 inhibitors (803) and NF-κB was shown to mediate mRNA expression of the transporter CAT-2B (804). Thus, it is conceivable that these post-translational mechanisms of control also operate in HZ + IFN-γ-dependent Mϕ iNOS regulation.

Because of the dual role proposed for NO in malaria, it is difficult to speculate at this point whether the contribution of HZ to increase its production is harmful or beneficial to the host. Before drawing any conclusions, different points should be taken into account. First, our results cannot be extrapolated to all types of Mϕ because responses to HZ vary according to the Mϕ population. For instance, whereas HZ was shown to exert a down-regulating effect on NO production by peritoneal Mϕ, it did not modulate this activity in microglial cells (204) and it was able to increase it in bone marrow-derived Mϕ (see Figure 1) (774). In order to clarify this issue, the effects of HZ on IFN-γ-mediated NO production must be monitored in Mϕ from different tissue sources as well as in monocytes. Second, the ability of HZ to induce or enhance NO synthesis must be assessed *in vivo*. Given that NO production and its effects in malaria vary according to its organ distribution, phase of infection, parasite stage, and host/parasite strain (38, 59, 141, 144, 782, 785, 805, 806), when examining it in response to HZ as well as its effects on the host, all these variables should be considered. For instance, whereas contradictory results have been reported regarding a correlation between serum nitrate levels and human severe malaria (142, 143), high levels of iNOS expression were detected in brain tissue from fatal cases of CM (144), underlining the importance of monitoring local iNOS and NO production during CM. Thus, as a first step towards defining the role of HZ in the development of CM, HZ accumulation and its ability to increase iNOS expression and NO synthesis must be examined in the brain. Third, even though a role for NO has been proposed in human CM (141-144, 781), and iNOS expression in circulating monocytes has been correlated with increased systemic NO production in malaria infected children (143), the ability of human Mϕ to produce NO is still controversial (481-485). Therefore, whether HZ enhances human monocyte/Mϕ NO generation in response to IFN-γ needs to be clarified. Altogether, these future studies should contribute to better define the participation of HZ in NO regulation during both human and rodent malaria.

Plasmodium, as well as a diverse range of parasites (*e.g., Leishmania, Toxoplasma, Trypanosoma*) interacts with the chemokine–receptor network. Such interactions contribute to determine the parasite-host relationship because of the different roles that chemokines and their receptors can play during infection, including microbicidal actions and detrimental effects to the host (615). On the one hand, the chemokine system has the crucial role of linking innate and adaptive immune responses via leukocyte recruitment to sites of infection and APC migration to lymph nodes to activate naïve T and B cells (599). However, high expression of chemokines and/or their receptors can also create the potential for persistent or inappropriately exacerbated response (*e.g.,* tissue damage, microvascular obstruction). In most cases, temporal and spatial patterns of chemokine/receptor expression are interwoven with ligand-receptor specificity to result in a multicomponent pathophysiological response (612). Therefore, in the second part of our studies, we were interested to identify the members of the chemokine system induced in response to HZ and to unravel the mechanisms controlling their expression in the Mϕ, one of the major sources of inflammatory mediators in malaria (95). A better understanding of how parasite components modulate this part of the immune response will be helpful to gain some insight into the pathogenesis of severe malaria.

As depicted in Figure 4, we found that HZ increases the expression of various Mϕ chemokine transcripts (MIP-1α, MIP-1β, MIP-2 and MCP-1) (776). Chemokine mRNA up-regulation was detected at early times of HZ stimulation and occurred in a concentration-dependent manner, suggesting that upon its release into the microvasculature, HZ might interact with resident Mϕ and lead to rapid chemokine production. In this context, HZ could stimulate the immunological response in the sites of parasite sequestration by inducing potent chemoattractants and activators. However, at higher parasitemias and subsequent higher HZ concentrations, the malaria pigment might also contribute to an exacerbated proinflammatory response (*e.g.,* leukocytosis, high ROS production) causing tissue damage and microvascular flow disturbance. The increase in HZ-inducible chemokine mRNA levels occurred through a mechanism dependent on MEK1/2-ERK1/2- and ROS- signals (Figure 5 and Figure 6, respectively) (776). The intracellular events triggered by these second messengers resulted in the reduction of Mϕ PTP activity, the nuclear translocation of NF-κB and its subsequent binding to the various chemokine promoters. Despite the activation of several transduction signals, this cellular regulation was specific, since other kinases -including p38 and JNK MAPK- shown to be involved in chemokine regulation were not activated in response to HZ.

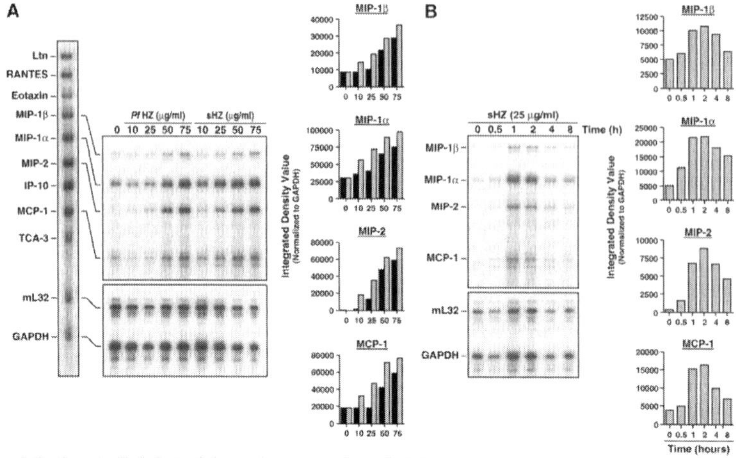

Figure 4. *P. falciparum* HZ and sHZ increase chemokine mRNA levels in murine Mϕ. Cells were treated either with 10-75 μg/ml *Pf*HZ or sHZ for 2 h (A) or with 25 μg/ml sHZ over an 8-h period (B) and chemokine mRNA expression was monitored using a mCK5 multi-probe RPA system (*left panels*). Densitometric quantification of chemokine mRNA levels over negative control after normalization to GAPDH (*right panels*). *Pf*HZ (solid bars), sHZ (gray bars). Reprinted with permission, from *The Journal of Immunology*. Vol 174:475-484. Jaramillo, M., M. Godbout, and M. Olivier. Hemozoin induces macrophage chemokine expression through oxidative stress-dependent and -independent mechanisms. ***Copyright 2005. The American Association of Immunologists, Inc***. Reference (776).

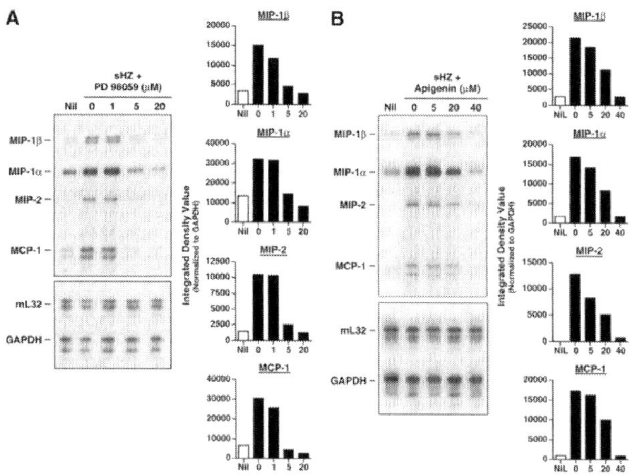

Figure 5. The ERK1/2 pathway is involved in chemokine mRNA induction in response to sHZ. Mϕ were incubated for 1 h with PD 98059 (A) or apigenin (B) before sHZ-stimulation (2 h) and their effect on chemokine mRNA expression was evaluated by RPA (*left panels*). Integrated density values of chemokine mRNA levels normalized to GAPDH (*right panels*). Nil: untreated (open bars), sHZ ± PD 98059/apigenin (solid bars). Reprinted with permission, from *The Journal of Immunology*. Vol 174:475-484. Jaramillo, M., M. Godbout, and M. Olivier. Hemozoin induces macrophage chemokine expression through oxidative stress-dependent and -independent mechanisms. ***Copyright 2005. The American Association of Immunologists, Inc***. Reference (776).

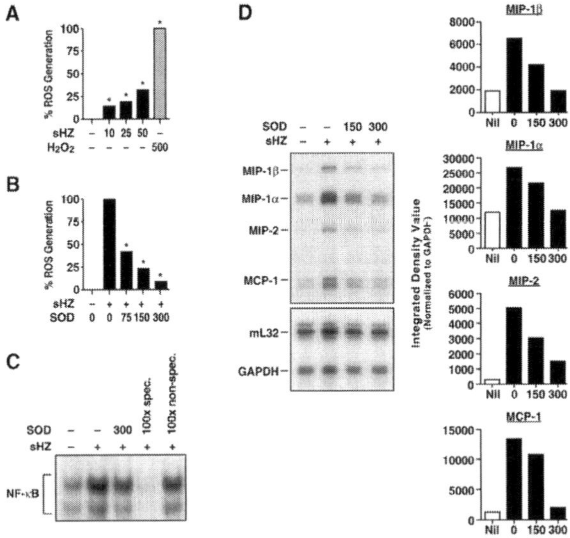

Figure 6. HZ-mediated chemokine induction is dependent on ROS generation. (A) Cells were stimulated with either 10-50 μg/ml sHZ or 500 μM H_2O_2 for 1 h and ROS generation was measured. sHZ (solid bars), H_2O_2 (gray bars). *, $p < 0.05$, sHZ or H_2O_2 vs Nil (OD measurements). (B) Mϕ pretreated or not with 0-300 U/ml SOD for 1 h were incubated with sHZ and ROS generation was monitored after 1 h. *, $p < 0.05$, sHZ + SOD vs sHZ (OD measurements). (C) Cells were stimulated as described in (B) but this time nuclear proteins were extracted and subjected to EMSA. (D) Following Mϕ stimulation with sHZ (2 h) either in absence or in presence of 0-300 U/ml SOD, total RNA was extracted and chemokine mRNA modulation was evaluated by RPA. Nil (open bars), sHZ ± SOD (solid bars). Reprinted with permission, from *The Journal of Immunology*. Vol 174:475-484. Jaramillo, M., M. Godbout, and M. Olivier. Hemozoin induces macrophage chemokine expression through oxidative stress-dependent and -independent mechanisms. ***Copyright 2005. The American Association of Immunologists, Inc***. Reference (776).

Our data indicating the requirement for MEK1/2-ERK1/2 and NF-κB activation are in agreement with several studies showing their involvement in chemokine induction by pathogens (*e.g., H. pylori, Trichomonas vaginalis)* (690, 699), cytokines (*e.g.,* TNF-α) (684-686) and host inflammatory particles (*e.g.,* uric acid, MSU crystals) (703-705) (See Chapter 4, Table 2. Signaling Events Involved in Chemokine Gene Expression). Of interest, in addition to ERK1/2 and NF-κB, the involvement of ROS was also reported (685, 686, 705), which might be related to their ability to act as a "danger signal" to the immune response (599). Because an important role of chemokines is to alert the immune system against pathogen invasion or tissue injury (612), it is plausible that immune cells have developed a rapid response to infection or inflammation through the activation of common intracellular messengers. Further supporting this idea, we found that the up-regulating effect of HZ on IFN-γ-mediated NO production also occurred via ERK1/2-dependent NF-κB activity. Thus, it is likely that by triggering the same intracellular signals, HZ modulates different Mϕ functions, including chemokine expression and amplification of Mϕ responses to IFN-γ (*e.g.*, NO), which would in turn contribute to enhance the inflammatory response during malaria infection.

We showed that HZ-inducible NF-κB activity was, at least in part, mediated by ERK1/2-dependnet signals (Figure 7A) (776). Consistent with this, previous studies have demonstrated that either specific inhibitors (807-809) or constitutive mutants (810) of the MEK1/2-ERK1/2 pathway abrogated NF-κB activation. Although the mechanism through which ERK1/2 enhances NF-κB translocation needs to be investigated, it is known that Ras/Raf/MEK1/2-ERK1/2-mediated signals can lead to NF-κB activation by phosphorylating and activating its positive regulators, IKKs (725, 736, 811). Indeed, Ras/Raf-activated MEKK1 phosphorylates subunits of the IKK complex and modulates their activity (725). Even though MEKK1 can activate both IKKα and IKKβ (811), it preferentially phosphorylates IKKβ on Ser177 and Ser181 (736). Whether MEKK1-mediated IKK phosphorylation constitutes a key step in HZ-inducible NF-κB nuclear translocation and chemokine transcription remains to be elucidated.

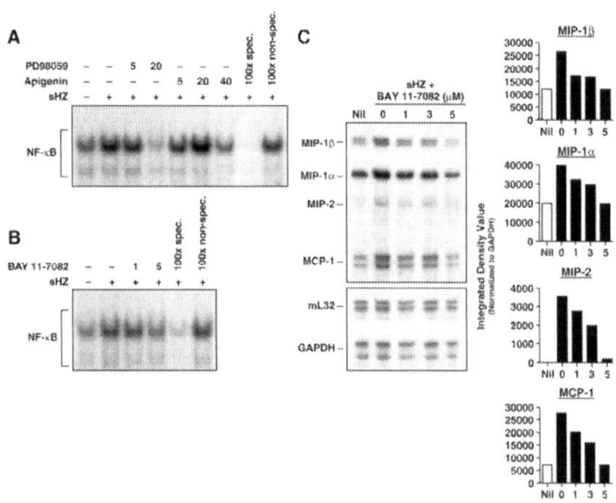

Figure 7. Role of NF-κB on Mϕ sHZ-dependent chemokine mRNA up-regulation. Labeled NF-κB probe was incubated with nuclear extracts from either untreated or sHZ-stimulated cells (1 h) pretreated or not with (μM) of apigenin, PD 98059 (A) or BAY 11-7082, a NF-κB inhibitor (B) and EMSA analyses were performed. (C) After a 1-h-exposure to BAY 11-7082, Mϕ were further stimulated with sHZ for 2 h. Then, total RNA was extracted and changes on chemokine mRNA levels were monitored by RPA. Nil: untreated (open bars), sHZ ± BAY 11-7082 (solid bars). Reprinted with permission, from *The Journal of Immunology*. Vol 174:475-484. Jaramillo, M., M. Godbout, and M. Olivier. Hemozoin induces macrophage chemokine expression through oxidative stress-dependent and -independent mechanisms. ***Copyright 2005. The American Association of Immunologists, Inc***. Reference (776).

In addition to the involvement of the MEK1/2-ERK1/2 pathway, we found that NF-κB nuclear translocation was partially dependent on ROS generation by HZ (see Figure 6C) (776). These observations are consistent with our data demonstrating the activation of this transcription factor in response to oxidative stress, more specifically to H_2O_2 (Figure 8) (777). In this study, we showed that H_2O_2 activated MEK1/2-ERK1/2-dependent signals, which were in turn required for NF-κB activation and chemokine gene expression (Figure 9) (777). However, ERK1/2 phosphorylation by HZ appeared to be independent of ROS generation, since SOD pretreatment did not affect this event (data not shown). These discrepancies might stem from the fact that the oxidative stress produced in response to HZ was much lower than the one generated by exogenous 500 μM H_2O_2 (see Figure 6A) (776), the concentration used in our ROS-induced chemokine work (777). Thus, it is plausible that different levels of oxidative stress trigger different Mϕ signaling pathways.

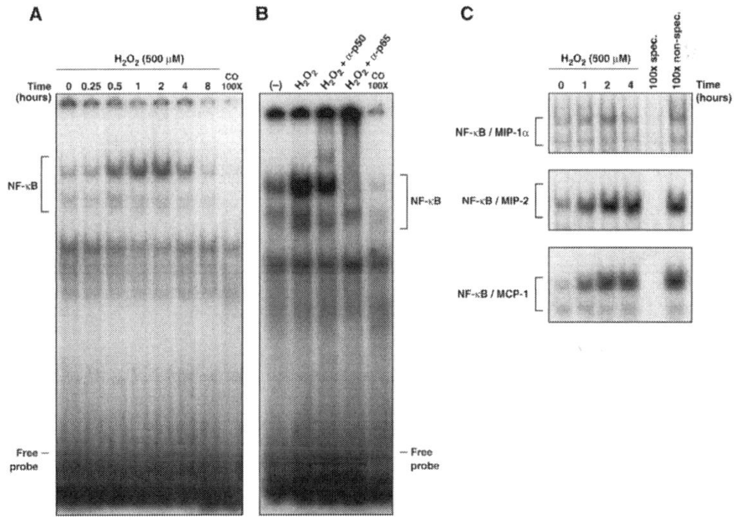

Figure 8. H_2O_2 induces NF-κB nuclear translocation and binding to several chemokine promoters in murine Mϕ. (A) Nuclear extracts from cells stimulated with H_2O_2 for different time periods (0-8 h) were subjected to EMSA. (B) For supershift assays, nuclear extracts from cells treated with H_2O_2 (2 h) were incubated or not with specific Abs against the p50 and p65 NF-κB isoforms for 1 h before EMSA. (C) EMSA analysis was performed as described in (A), but this time nuclear extracts were incubated with one of three NF-κB probes specific for the murine MIP-1α, MIP-2 or MCP-1 promoters. Reprinted with permission, from *The Journal of Immunology*. Vol 169:7026-7038. Jaramillo, M., and M. Olivier. Hydrogen peroxide induces murine macrophage chemokine gene transcription via extracellular signal-regulated kinase- and cyclic adenosine 5'-monophosphate (cAMP)-dependent pathways: involvement of NF-kappa B, activator protein 1, and cAMP response element binding protein. ***Copyright 2002. The American Association of Immunologists, Inc.*** Reference (777).

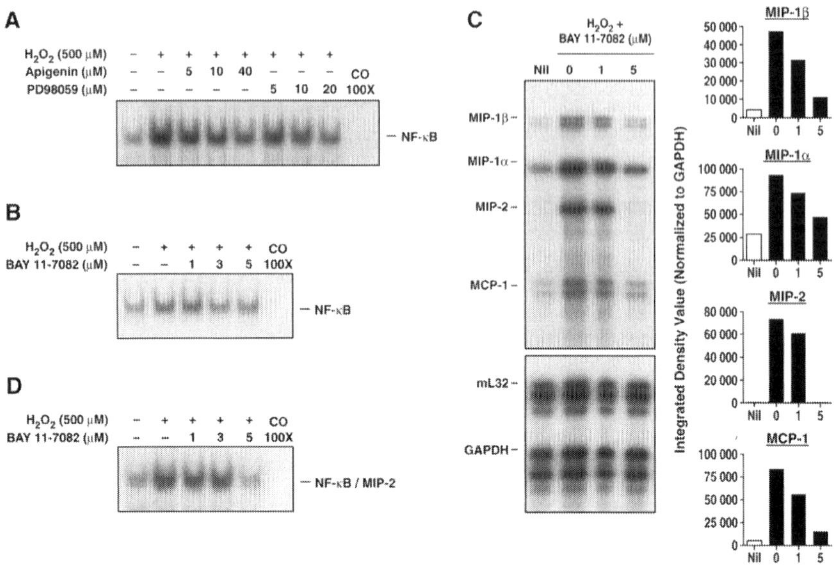

Figure 9. Role of NF-κB on Mϕ chemokine mRNA expression in response to H_2O_2. Nuclear extracts from either untreated or H_2O_2-stimulated cells (2 h) pretreated or not with apigenin or PD 98059 (1 h) were incubated with a labeled NF-κB oligonucleotide and subjected to EMSA (A). Either a consensus NF-κB probe (B) or a probe containing a NF-κB binding site in the murine MIP-2 promoter (D) were incubated with nuclear proteins from cells treated with BAY 11-7082 (1 h) before H_2O_2 stimulation (2 h), and EMSA were performed. Total RNA was extracted from Mϕ stimulated as described in the previous experiment, and chemokine mRNA modulation was monitored by RPA (C). Nil: untreated (open bars), H_2O_2 ± BAY 11-7082 (solid bars). Reprinted with permission, from *The Journal of Immunology*. Vol 169:7026-7038. Jaramillo, M., and M. Olivier. Hydrogen peroxide induces murine macrophage chemokine gene transcription via extracellular signal-regulated kinase- and cyclic adenosine 5'-monophosphate (cAMP)-dependent pathways: involvement of NF-kappa B, activator protein 1, and cAMP response element binding protein. ***Copyright 2002. The American Association of Immunologists, Inc***. Reference (777).

As to the mechanisms through which HZ-inducible ROS activate NF-κB and lead to chemokine gene expression, our data showed the requirement of Mϕ PTP down-regulation with the subsequent increase of PTK activity (Figure 10) (776). Consistent with this proposal, we previously showed enhanced NF-κB activity in response to specific PTP inhibitors (812) and others reported that SOD pretreatment abrogated Mϕ IκBα tyrosine phosphorylation (813) and the activation of IKKα, IKKβ and the p65 NF-κB subunit (814). Of interest, our *in vitro* data revealed that bpV (phen), a potent and selective PTP inhibitor (815), suppressed basal Mϕ phosphatase activity (816) and increased chemokine mRNA levels in Mϕ (817). In parallel, we detected a greater level of chemokine induction in Mϕ deficient in SHP-1 -one of the major Mϕ PTPs- when compared to wild type Mϕ (817). Moreover, we demonstrated that PTP inhibition strongly up-regulated the *in vivo* expression of various chemokine transcripts and enhanced their production in response to infection (452). In light of these findings, it seems plausible to propose ROS-mediated PTP inactivation as a possible mechanism for HZ-inducible NF-κB activity and Mϕ chemokine modulation. From these results derives the important issue of identifying the signaling events involved in this cellular regulation. This can be addressed by evaluating: i) the phosphorylation of IκBα on tyrosine 42, which occurs in response to oxidative stress (818); ii) the HZ-mediated phosphorylation of specific PTK(s), such as c-Src and Syk tyrosine kinases, reported to be involved in IκB tyrosine phosphorylation and NF-κB activation by ROS (819, 820); iii) the activation of members of the IKK complex by HZ and the upstream events leading to such regulation (*e.g.,* MEKK1 and NIK activation) (725); iv) the serine phosphorylation of p65, shown to be induced by oxidative stress (820); and v) the identification of the specific PTP(s) that are inactivated in response to HZ and to evaluate the impact of these events on IKK, IκB and ultimately, NF-κB and chemokine regulation.

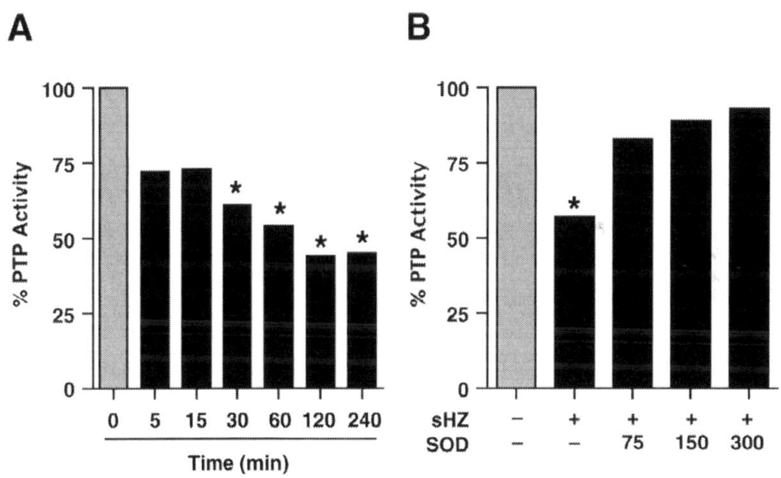

Figure 10. SOD restores Mϕ PTP inactivation in response to sHZ (A) Protein lysates from Mϕ either left untreated or stimulated with 25 μg/ml sHZ over a 4-h period were subjected to PTP activity assay (pNPP hydrolysis). *, $p < 0.05$, sHZ vs Nil (OD measurements). (B) Following 0-300 U/ml SOD treatment cells were stimulated with sHZ for 1 h and PTP activity was monitored. *, $p < 0.05$, sHZ + SOD vs sHZ (OD measurements). Nil: untreated (gray bars), sHZ ± SOD (solid bars). Reprinted with permission, from *The Journal of Immunology*. Vol 174:475-484. Jaramillo, M., M. Godbout, and M. Olivier. Hemozoin induces macrophage chemokine expression through oxidative stress-dependent and -independent mechanisms. ***Copyright 2005. The American Association of Immunologists, Inc***. Reference (776).

Besides ROS- and ERK/2-dependent signals, we did not examine the potential contribution of other pathways to HZ-mediated chemokine modulation. However, our results revealed that multiple chemokine mRNA up-regulation by H_2O_2 required the cooperation between ERK1/2- and cAMP-dependent signals leading to the formation of a transcriptional complex (NF-κB, AP-1 and CREB) (Figure 11) (777). These findings are in line with several reports showing the cooperative binding of various regulatory elements forming an enhanceosome (transcription factors, coactivators), which appears to be essential for maximal activation of some chemokine promoters (*e.g* MCP-1, RANTES, IL-8) (669, 670, 677, 678). The requirement of H_2O_2-inducible ERK1/2 in this regulatory process is in agreement with their ability to control multiple transcription factor activation at different levels. Regulation of c-Fos (a member of the AP-1 family) occurs at the transcriptional level and involves ERK1/2-dependent CREB phosphorylation (723), which then binds to the CRE in the *c-fos* promoter (742, 743). c-Fos can form heterodimers with c-Jun and bind the AP-1 site on the *c-jun* promoter region inducing its transcription (744). Moreover, ERK1/2 can directly phosphorylate c-Jun at various serine residues and increase AP-1 activity (720, 746). In addition, our data suggesting a crosstalk between cAMP and ERK1/2 are consistent with studies demonstrating that cAMP is able to potentiate ERK1/2 phosphorylation (712, 821) and to increase AP-1 expression via the CRE present in the promoter of *c-fos* (822). Altogether, these data indicate that chemokine gene expression might be finely regulated through the interplay of diverse signaling pathways leading to the cooperative actions of transcription factors on the chemokine promoters.

Because ROS and ERK1/2 were found to be involved in chemokine induction by HZ, and our preliminary data indicate that, in addition to NF-κB, HZ activates other transcription factors (*e.g.*, AP-1, CREB) (data not shown), it is conceivable that complementary pathways contribute to the formation of a transcriptional complex required for maximal chemokine gene expression in response to HZ. Figure 12 depicts the signaling events until now identified in HZ-dependent Mϕ chemokine mRNA expression. However, further investigation of the various kinases and transcription factors, and a better understanding of their protein-protein and protein-DNA interactions is required in order to fully unravel this regulatory process. Although the Mϕ receptor(s) involved are still unknown, several lines of evidence point to a role for members of the TLR family. First, professional phagocytes such as Mϕ express TLRs, which function as pattern-recognition receptors that recognize conserved molecular patterns shared by a large group of microorganisms, and are essential for translating the recognition of microbial components to activation of the immune system (644). Thus, it is possible that a repetitive structure such as that of HZ crystals be recognized via TLRs. Second, it is now widely accepted that chemokine induction occurs directly in response to pathogens or pathogenic components via TLR ligation and triggering of signal transduction (598, 643) (See Chapter 4, – Table 1, Chemokine Induction *via* TLR Activation). Third, activation of TLRs (*e.g.,* TLR-2, TLR-4) in response to parasitic infections, including *T. cruzi* (75, 659), *T. gondii* (76) and *L. major* (77, 658), appears to play an important role in host defense. Although this mechanism of immunity remains largely unexplored in malaria, two separate studies have reported TLR-mediated signaling in response to *Plasmodium.* In mice deficient in MyD88 (an adaptor molecule shared by TLRs) *P. berghei* infection did not result in elevation of IL-12 levels and did not cause liver injury (78). Moreover, a recent study demonstrated that *P. falciparum* activates DCs via TLR9-dependent signaling (79). These data suggest that the TLR system is also important in the inflammatory response during malaria infection. Future investigation of TLR involvement in Mϕ activation by HZ as well as their role in HZ-inducible inflammatory mediator *in vivo* will bring light on this important matter.

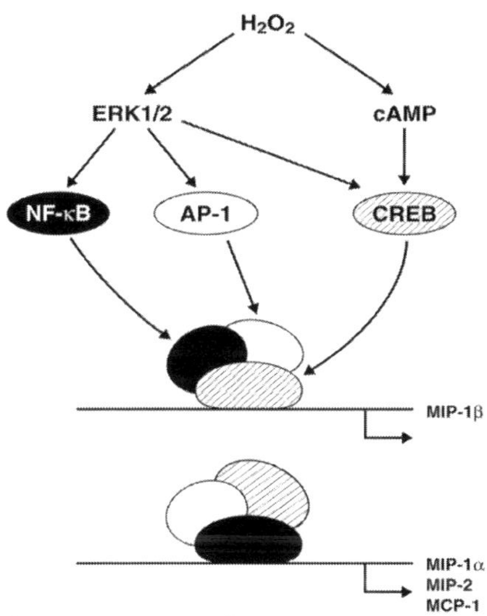

Figure 11. Hypothetical model of H_2O_2-mediated chemokine transcriptional activation in murine Mϕ. Following cell exposure to H_2O_2, ERK- and cAMP-dependent signaling pathways are activated, leading to the nuclear translocation and DNA binding of NF-κB, AP-1 and CREB. These transcription factors seem to form a multiprotein complex that might adapt its spatial conformation in order to bind to the available specific binding sites present in the murine MIP-1β, MIP-1α, MIP-2 and MCP-1 promoters and in such a way lead to maximal Mϕ chemokine transcription activation. Reprinted with permission, from *The Journal of Immunology*. Vol 169:7026-7038. Jaramillo, M., and M. Olivier. Hydrogen peroxide induces murine macrophage chemokine gene transcription via extracellular signal-regulated kinase- and cyclic adenosine 5'-monophosphate (cAMP)-dependent pathways: involvement of NF-kappa B, activator protein 1, and cAMP response element binding protein. ***Copyright 2002. The American Association of Immunologists, Inc.*** Reference (777).

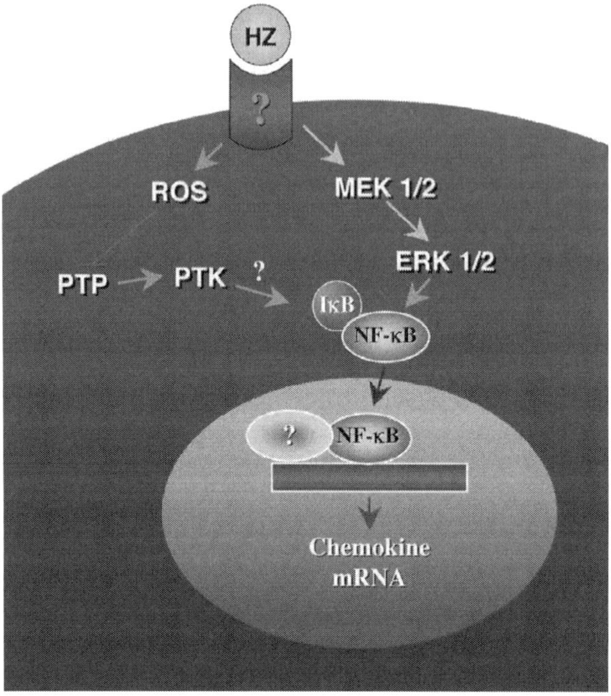

Figure 12. Hypothetical Model of Hemozoin-dependent Mϕ Chemokine Induction. Reprinted with permission from Landes Bioscience. Olivier, M. and M. Jaramillo. Modulation of positive signaling and proinflammatory responses by hemozoin, a *Plasmodium* metabolic waste. In: Denkers, E. Y. and R. T. Gazzinelli, eds. Protozoans in Macrophages. Austin. Landes Bioscience. 2007: 67-83.

In addition to transcriptional regulation, increased chemokine mRNA levels in response to pathogens or cytokines may be the result of post-transcriptional mechanisms of control (See Chapter 4, – Table 3, Post-transcriptional Regulation of Chemokines). These regulatory events appear to be related to the presence of AREs in the 3' -UTRs of the chemokine mRNAs, which are implicated in mRNA destabilization (747). AREs have been identified in MIP-1α, MCP-1 and MIP-2 mRNAs (668, 748, 750-752). Even though the mechanisms responsible for the increase in chemokine mRNA $t_{1/2}$ are not yet completely understood, members of the MAPK family (including ERK1/2) have been shown to enhance the stability of certain chemokine transcripts that bear one or more AREs (*e.g.,* MIP-2, IL-8) (754, 755). Besides ERK1/2, we (777) and others (751, 769) have reported that ROS also have the ability to increase chemokine mRNA $t_{1/2}$ (Figure 13B) (777). Thus, it is possible that through ROS generation and/or ERK1/2 activation, HZ controls Mϕ chemokine mRNA expression at the post-transcriptional level.

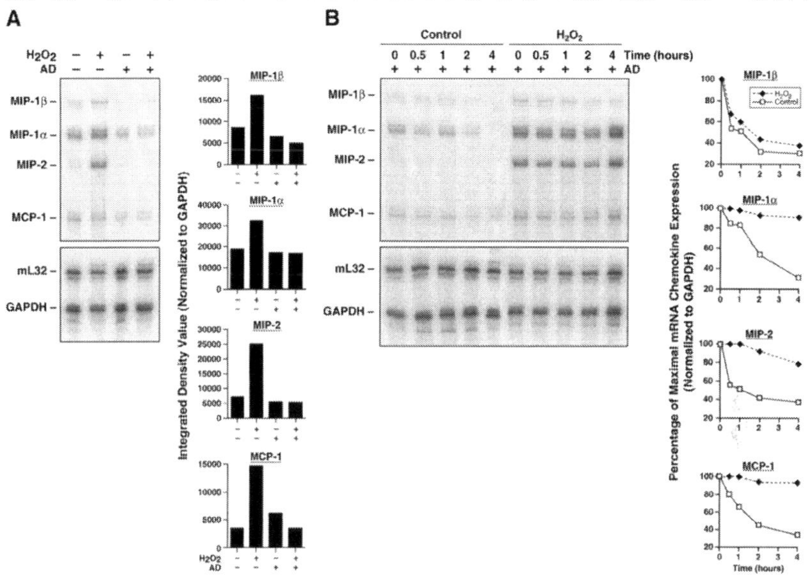

Figure 13. Transcriptional and post-transcriptional regulation of Mϕ chemokines by H_2O_2. (A) B10R cells were stimulated with H_2O_2 for 2 h either in absence or in presence of 5 μg/ml AD. Next, total RNA was extracted and chemokine mRNA levels were monitored by RPA (*left panel*). Densitometric quantification of chemokine mRNA induction over negative control after normalization to GAPDH (*right panel*). (B) Cells were left untreated (control) or stimulated with H_2O_2 for 2 h. Then, AD was added for different time periods (0-4 h). Following total RNA isolation, RPA analysis was performed (*left panel*). Densitometric quantification of the decay of chemokine mRNA normalized to GAPDH (*right panel*). H_2O_2 (solid diamonds), control (open squares). Reprinted with permission, from *The Journal of Immunology*. Vol 169:7026-7038. Jaramillo, M., and M. Olivier. Hydrogen peroxide induces murine macrophage chemokine gene transcription via extracellular signal-regulated kinase- and cyclic adenosine 5'-monophosphate (cAMP)-dependent pathways: involvement of NF-kappa B, activator protein 1, and cAMP response element binding protein. ***Copyright 2002. The American Association of Immunologists, Inc.*** Reference (777).

In contrast to previous reports associating HZ-inducible ROS to the inhibition of various Mϕ functions (*e.g.,* phagocytosis, LPS-mediated cytokine and NO production, PMA-inducible ROS) (67, 204), we found that ROS generation is one of the mechanisms through which HZ leads to Mϕ chemokine induction (see Figure 6) (776). As previously discussed (204), these apparently contradictory results are likely to be related to the differential susceptibility of various Mϕ populations to oxidative stress. Investigation of Mϕ susceptibility to the pro-oxidant activity of HZ revealed that the extent of lipid peroxidation was higher in peritoneal Mϕ than in J774 murine Mϕ and microglial cells; an event associated with differences in their membrane lipid content and/or their GSH levels (421). These data offer a new interpretation for the contentious immunomodulatory effects of HZ reported for phagocytes of different origins. In fact, whereas inhibitory effects were detected in peritoneal Mϕ (121, 196, 197, 204), cellular activation was observed in various Mϕ cell lines by others (61) and by us (774, 776), and no modulation occurred in microglial cells (204). In this context, it is conceivable that the effects of HZ can vary according to the cell type and its tissue source in association with its antioxidant defenses. In support of this idea is the fact that cells are generally subjected to a redox regulation, pro-oxidant in the cytosol and reductant in the nucleus (823), and since the equilibrium between oxidants and antioxidants within these compartments appears to be cell type specific (824), the same signaling pathway may act in opposite ways in different cells. For instance, whereas ERK1/2 MAPK acted as negative regulators of NF-κB in H_2O_2-treated epithelial cells (825), we obtained the opposite results in bone marrow-derived Mϕ (see Figure 9A) (777). Thus, it will be of paramount importance to evaluate the up-regulating versus suppressing actions of HZ in different monocytes/Mϕ populations and cell types, and to determine whether such effects contribute to malaria immunopathology. Because of their activation in malaria and their important functions during inflammation and host defense, special attention should be given to the evaluation of HZ effects in Nϕ and DCs.

Nϕ-mediated parasite killing has been reported in response to *P. falciparum* infection (826, 827). The microbicidal mechanisms include increased phagocytic activity (773), ROS generation (828, 829), and elastase production (830). Of interest, the concentrations of Nϕ elastase correlated with disease severity in infected patients, suggesting that Nϕ activation may play a role in the pathogenesis of severe malaria (830). Further supporting this hypothesis, experimental evidence indicates that Nϕ contribute to the pathophysiology of ECM. Indeed, Nϕ depletion in circulation prevented early development of ECM and dramatically decreased microhemorrhage and monocyte sequestration in the brain. Moreover, abrogation of IL-12 mRNA expression and down-regulation of TNF-α, IFN-γ and IL-2 transcripts in the brain was also detected (165, 166). In correlation with these data, Nϕ from mice suffering from CM expressed various cytokine (IL-2, IL-18, IFN-γ, TNF-α) and chemokine (MIP-1α, IP-10, Mig) transcripts (167), suggesting that Nϕ might participate in the pathogenesis of ECM via proinflammatory cytokine and chemokine expression.

Whether HZ is, at least in part, responsible for Nϕ activation in malaria remains to be established; however, some indirect evidence, along with a few experimental data, points to an interaction between the malaria pigment and these immune cells. HZ phagocytosis and accumulation has been detected in Nϕ (198, 386-388). Moreover, intraleukocytic pigment was associated with cellular activation and disease severity. Among adult patients with severe *P. falciparum* malaria, those who succumbed to infection had significantly higher proportions of HZ-containing Nϕ on admission than did survivors (386). In addition, the count of peripheral HZ-containing Nϕ was a better indicator of bad prognosis than the peripheral parasite count in both adults (386) and children (388). Despite these observations, very little attention has been paid to the evaluation of HZ actions in Nϕ. Comparison of the effects of HZ in monocytes and Nϕ revealed that whereas PMA-induced ROS was reduced in HZ-laden monocytes, it remained unaffected in HZ-loaded Nϕ (198), indicating that the immunomodulatory properties of HZ are cell-type specific, and that the Nϕ is more resistant to the pro-oxidant effects of HZ. In fact, the ability of iRBC to trigger Nϕ ROS was much lower in iRBC containing ring forms than in those with mature forms of the parasite (828); thus, it is possible that HZ, a unique schizont component, participates in Nϕ activation via ROS generation. Further supporting an up-regulating effect of HZ in these cells, exposure of Nϕ to isolated *P. falciparum* HZ resulted in a significant increase of adhesion molecule expression (194). Through this mechanism, HZ could promote leukocyte adherence to vascular endothelium *in vivo*, thereby contributing to tissue injury and microvascular flow disturbance observed in malaria. Altogether, these studies suggest that the malarial pigment interacts with the Nϕ population and underline the importance of exploring the role of HZ in their activation during malaria.

In addition to Mϕ and Nϕ, DC activation also appears to be a feature in malaria infection (68, 69, 79). Incubation of DCs with *P. chabaudi* schizonts resulted in IL-6 and IL-12 synthesis as well as in increased expression of maturation markers (CD86) and MHC-II molecules (68). In line with these findings, it was recently shown that *P. falciparum* activates human and murine DCs (*e.g.,* CD86 up-regulation, IFN-γ production) *via* TLR9-MyD88-dependent signaling (79). That HZ is involved in DC maturation is strongly suggested by a study performed by Coban *et al.* (69). When immature human DCs were incubated with *P. falciparum* HZ, significant increases in the maturation markers for DCs (CD83, CD86, and CD1a) were detected. Moreover, the production of IL-12 and the morphological features characteristic of mature DC were also observed. DCs are crucial APCs (435), and they also activate innate immune responses to microbial challenge by producing chemokines (601) and cytokines (436). Thus, the interaction between DC and parasite components could result in the activation of an adaptive immune response, but also in inflammatory mediator-associated pathology. Further examination of the effects of HZ on these important DC functions will be helpful to better define the immunomodulatory properties of this parasite metabolite.

Despite the demonstration that HZ induces cytokine and chemokine expression *in vitro* (61, 65, 205), a causal relationship between HZ and proinflammatory mediator release had not been established *in vivo*. Thus, we next focused on the clarification of this important issue. In the first part of our study, we made use of the murine air pouch system, an experimental approach commonly used (148, 576), because it provides a suitable closed environment to evaluate multiple inflammatory parameters in response to a given stimulus. HZ injection in the pouch resulted in significant leukocyte recruitment (Figure 14), which was accompanied by increased levels of various chemokines (MIP-1α, MIP-1β, MIP-2/CXCL2 and MCP-1) (Figure 15), chemokine receptors, (CCR1, CCR2, CCR5, CXCR2, CXCR4), proinflammatory cytokines (IL-1 and IL-6) and MRPs (S100A8, S100A9 and S100A8/A9) (775). Based on these results, and in order to mimic HZ release into the circulation during malaria, HZ was next inoculated i.v. The effects of HZ were monitored in the liver since its accumulation in this organ has been associated with infection chronicity in humans (393), and with cumulative parasite burden and duration of infection in experimental malaria (399). We found that HZ caused a concentration-dependent increase of various chemokine (MIP-1α, MIP-1β, MIP-2 and MCP-1) and cytokine (IL-1β and IL-12) transcripts in the liver (Figure 16).

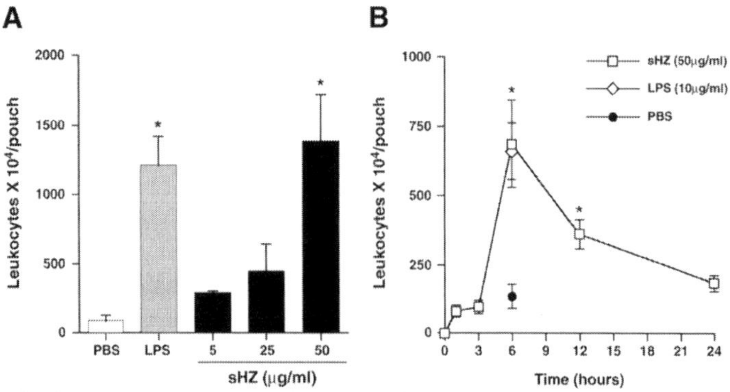

Figure 14. sHZ induces leukocyte recruitment *in vivo*. (A) Number of leukocytes accumulating in the air pouch of BALB/c mice in response to 5-50 μg/ml sHZ, 10 μg/ml LPS or PBS after 6 h. PBS (open bars), LPS (gray bars), sHZ (solid bars). *, $p < 0.05$, sHZ or LPS vs PBS. (B) Kinetics of total leukocyte recruitment in response to 50 μg/ml sHZ over a 24-h period. *, $p < 0.05$, sHZ vs PBS. Results represent mean + SEM of five mice. Reprinted with permission, from *The Journal of Immunology*. Vol 172:3101-3110. Jaramillo, M., I. Plante, N. Ouellet, K. Vandal, P. A. Tessier, and M. Olivier. Hemozoin-inducible proinflammatory events in vivo: potential role in malaria infection. ***Copyright 2004. The American Association of Immunologists, Inc***. Reference (775).

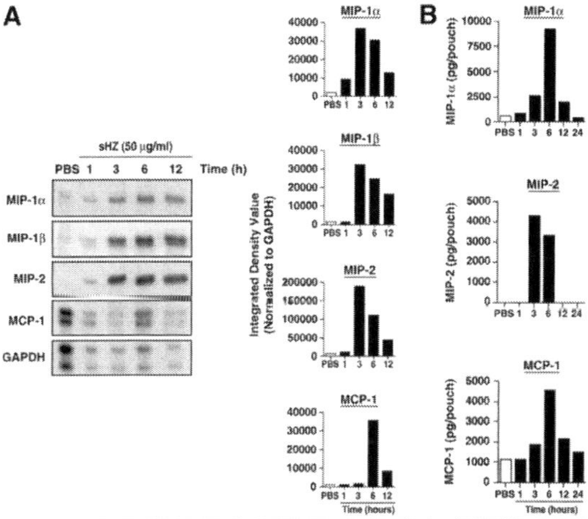

Figure 15. sHZ-mediated chemokine up-regulation in the air pouch. (A) Time course of chemokine mRNA following sHZ (50 μg/ml) inoculation over a 12-h period. Exudates from each experimental group were pooled, total RNA was extracted from the leukocytes recruited and samples were submitted to RPA (B) Kinetic analyses of MIP1α, MIP-2 and MCP-1 protein expression in the air pouches of BALB/c mice in response to sHZ. Supernatants of pooled exudates from 1- to 24-h-treated mice were subjected to ELISA. PBS (open bars), sHZ (solid bars). Data represent chemokine mRNA and protein expression in pooled samples (five mice) from each experimental group. Reprinted with permission, from *The Journal of Immunology*. Vol 172:3101-3110. Jaramillo, M., I. Plante, N. Ouellet, K. Vandal, P. A. Tessier, and M. Olivier. Hemozoin-inducible proinflammatory events in vivo: potential role in malaria infection. ***Copyright 2004. The American Association of Immunologists, Inc***. Reference (775).

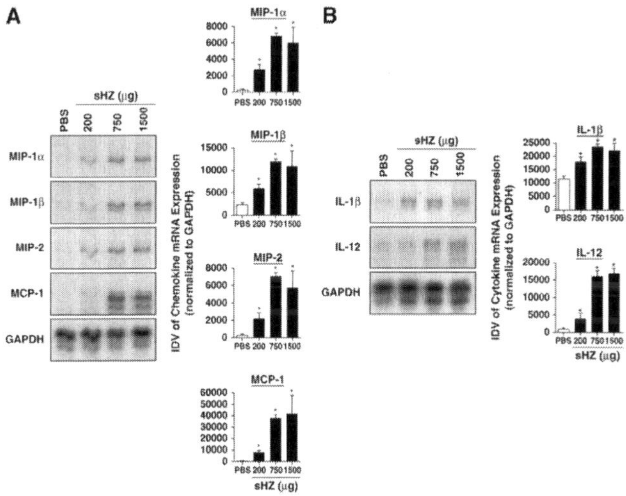

Figure 16. sHZ induces liver chemokine and cytokine mRNA expression. BALB/c mice were injected i.v. in the tail vein either with PBS or with sHZ. After a 6-h treatment, total RNA extraction from the liver was followed by RPA to monitor chemokine (A) and cytokine (B) mRNA expression. Data presented in *left panels* correspond to one animal representative of each experimental group. Graphs in *right panels* represent integrated density values of mean chemokine and cytokine mRNA levels normalized to GAPDH. Results represent mean + SEM of five mice. PBS (open bars), sHZ (solid bars). *, $p < 0.05$, sHZ vs PBS. Reprinted with permission, from *The Journal of Immunology*. Vol 172:3101-3110. Jaramillo, M., I. Plante, N. Ouellet, K. Vandal, P. A. Tessier, and M. Olivier. Hemozoin-inducible proinflammatory events in vivo: potential role in malaria infection. ***Copyright 2004. The American Association of Immunologists, Inc***. Reference (775).

Even though the cellular sources of these inflammatory mediators remain to be identified, several lines of evidence point to a role for Kupffer cells and monocytes via ROS generation. In a pilot study Sherry and colleagues detected HZ accumulation in Kupffer cells upon i.v. injection of the pigment (unpublished data cited in (61)). Moreover, liver chemokine induction (61, 169) as well as monocyte liver infiltration and increased O_2^- levels (831) were reported in *Plasmodium*-infected mice. In addition, it has been shown that the major sources of chemokines and ROS during liver disease are Kupffer cells, and the production of ROS by these cells enhances the secretion of cytokines and chemokines (832). Because HZ-inducible Mϕ chemokine expression was in part ROS-dependent (776), and monocyte chemoattractans as well as inflammatory cytokines were detected in the liver of HZ-injected mice (see Figure 16) (775) , it is possible that HZ triggers ROS-mediated cytokine and chemokine production by Kupffer cells, which would in turn favour monocyte infiltration and further oxygen radical generation and inflammatory mediator release. This inflammatory response could contribute to reduce parasite burden, but could also favor exacerbated phagocyte activation, resulting in hepatic damage, as observed in both human and murine malaria (134, 833). Indeed, it has been proposed that ROS may play a role in the pathogenesis of severe malaria by causing endothelial damage (92), and their production has been associated with disease severity in adults and children (58, 93, 834). Thus, it will be important to assess ability of HZ to induce ROS and chemokine production in Kupffer cells, and to define the contribution of ROS to the effects of HZ *in vivo*. As to the experimental approach, the use of available *Sod* knockout mice (835) might provide helpful information.

In addition to defining the contribution of HZ-derived ROS in liver-associated inflammation during malaria, the actions of HZ must also be monitored in the spleen. During the course of experimental malaria there is rapid increase in the number of mononuclear cells in this organ (836), which is accompanied by local increased cytokine release (*e.g.,* IL-12, TNF-α and IFN-γ) (68) and ROS generation (831). The finding of splenomegaly during acute malaria infections in humans, suggest that similar leukocyte recruitment occurs (837). Because HZ accumulation has been detected in the spleen (61, 398, 399), and co-localization of HZ and proinflammatory mediator production has been reported in this organ (196), it is plausible that HZ participates in this localized inflammatory response during acute malaria.

The involvement of cytokines (*e.g.,* IL-1, IL-6, TNF-α) in HZ-mediated inflammatory effects (*e.g.,* chemokines, NO) also deserves further investigation. These mediators can stimulate chemokine (684, 838, 839) and NO (528, 840) production, and their participation in malaria pathology has been extensively investigated. High levels of TNF-α, IL-6 and IL-1β are more frequently observed in patients suffering from severe malaria than in those suffering from mild disease or asymptomatically infected (95-98, 138, 841). Moreover, cytokine overproduction, in particular of TNF-α and IL-1β, appears to be involved in the pathogenesis of human (139) and murine (158, 842) CM. Even though we detected increased levels of IL-1 and IL-6 in HZ-treated mice (775), we did not assess the *in vivo* production of TNF-α, which we plan to address in the future. Despite this lack of *in vivo* evidence, HZ treatment resulted in TNF-α production by human monocytes (198, 205) and murine peritoneal Mϕ (196). In line with these reports, we detected increased TNF-α mRNA levels in bone marrow-derived Mϕ (data not shown). Because cytokine production in response to HZ may offer a means of amplifying the inflammatory response, it will be important to assess whether this is a mechanism of cellular activation by the malaria pigment. The use of specific inhibitors or blocking Abs against the various cytokines might be helpful to define their participation in the various immunomodulatory effects of HZ both *in vitro* and *in vivo*.

Our data with the air pouch model demonstrate that HZ is a potent inflammatory agent *in vivo*, and our i.v. experiments indicate that upon its release, HZ appears to have an inflammatory effect in the liver (see Figures 14-16) (775). However, it is increasingly recognized that severe malaria is a disorder that affects several tissues and organs (*e.g.*, brain, placenta) (125). Therefore, evaluation of the local *in vivo* inflammatory properties of HZ will contribute to better define its role in the various syndromes related to severe malaria, such as PM and CM. During PM, inflammatory cells such as monocytes frequently accumulate in the intervillous spaces of the placenta in addition to iRBC. Even though the presence of mononuclear phagocytes is important to control parasitemia, the inflammatory reaction leads to placental lesions associated with poor birth outcomes (187, 189). In *P. falciparum*-infected pregnant women, TNF-α and IL-8 were found to be produced only by placental Mϕ that contained HZ, and their increased levels were associated with intrauterine growth retardation during PM (190). A correlation between placental monocyte recruitment and elevated MIP-1α and MCP-1 expression was also observed. Of interest, in some of the HZ-laden maternal Mϕ, production of MIP-1β and MCP-1 was detected (186). Consistent with this, we found that HZ enhances CC chemokine production (MCP-1, MIP-1α, MIP-1β) and induces monocyte recruitment *in vivo* (see Figures 14-16) (775). Therefore, it is conceivable that HZ is implicated in poor birth outcomes associated with enhanced leukocyte recruitment into the placenta and local inflammation. However, a causal relation between HZ and monocyte placental infiltration must be established *in vivo*, and its ability to enhance inflammatory mediator production by placental Mϕ must be assessed *in vitro*. In addition, given that placental trophoblasts cells avidly phagocytose parasitic material, and predominantly HZ granules (843), it will be interesting to monitor the actions of HZ in these cells.

Beyond its impact on PM, the accumulation and effects of HZ in the placenta may have the potential to facilitate HIV-1 replication and vertical transmission. Because malaria and HIV-1 co-infections are common in pregnant women in sub-Saharan Africa (844), by increasing monocyte numbers in the placenta, HZ could contribute to creating a reservoir for Mϕ-tropic HIV-1 virus. Moreover, we detected increased *in vivo* mRNA expression of various chemokine receptors that play a role in HIV-1 infection (775). CCR5 is a coreceptor required for efficient HIV-1 cell entry (193) and is the main secondary receptor implicated in HIV-1 vertical transmission (845). Interestingly, *P. falciparum* infection was shown to enhance CCR5 mRNA expression on placental Mϕ of pregnant women (192). In addition to CCR5, CCR2 and CXCR4 also serve as cofactors to permit HIV-1 entry (846, 847). Therefore, the potential of HZ favoring HIV-1 replication and intrauterine transmission during malaria and HIV-1 co-infections deserves further investigation.

It is now widely recognized that the local production of proinflammatory mediators contributes to progression of cerebral pathology in both human (137-139) and experimental (158, 842) malaria. Despite this common feature, some differences have been detected between both pathologies. Of particular relevance is the fact that there is little, if any, inflammatory infiltrate in the brain in humans, whereas mainly monocytes with a few iRBC are sequestered in the brain during ECM (154, 848, 849). Therefore, careful attention should be paid to the analysis of the potential effects of HZ in both contexts. Hopefully, murine models (*e.g., P. berghei* ANKA-[BALB/c x C57BL/6]; *P. yoelii*-BALB/c-Swiss), in which sequestration of iRBC in the brain was reported (155-157), will offer useful tools to better understand, and eventually control CM, one of the major complications related to *P. falciparum* malaria.

Despite the differences between human CM and ECM, our *in vivo* studies, along with data from human malaria patients and experimental models, support a role for HZ in the development of cerebral pathology. Evidence from autopsy and histological studies showed that HZ accumulates in the reticuloendothelial system and as patient conditions deteriorate and CM occurs, brain capillaries are also filled with iRBC and HZ-laden monocytes (395). Similarly, in an ECM model, the amount of HZ in the brain was higher in mice that developed this complication, indicating that the levels of HZ in this organ may be associated with CM (399). In line with this hypothesis, HZ inoculation in the air pouch resulted in the increase of various chemoattractants (MRPs, MIP-1α, MCP-1), chemokine receptors (CCR5) and leukocyte populations (monocytes, Nϕ) that are thought to contribute to CM. We found that HZ induces the production of various MRPs (S100A8, S100A9 and S100A8/S100A9), which are potent Nϕ chemoattractants and activators (148, 850, 851). During CM, analysis of cellular activation in the brain revealed S100A8 and S100A9 expression in microglial cells (147). Because Nϕ play a critical role in the development of ECM (165) and HZ accumulation occurs in the brain, it is possible that HZ-dependent MRP secretion and subsequent Nϕ infiltration contributes to ECM. Besides MRP secretion, microglial cells seem to participate in ECM development via TNF-α and chemokine (MCP-1, IP-10 and RANTES) induction in the brain (168, 169). Similarly, peritoneal Nϕ and Mϕ from mice suffering from CM expressed various cytokine (IL-2, IL-18, IFN-γ, TNF-α) and chemokine (MIP-1α, IP-10, Mig) transcripts (167). This is in agreement with data from ECM showing increased mRNA levels MCP-1 in brain of mice suffering of CM (169). In addition to inflammatory mediator production, an important role for CCR5 has been strongly suggested in ECM. It was recently demonstrated that CCR5 (164) and not CCR2 (852) expression, on both brain-sequestered leukocytes and nonhematopoietic cells, was required for ECM development. This effect was attributed to increased brain-sequestration of cytotoxic $CD8^+$ T cells, a leukocyte subset critical for CM progression (162, 163). Altogether, these studies indicate that Nϕ, Mϕ and microglial cells may take part in the pathogenesis of ECM via proinflammatory mediator and/or chemokine receptor expression. The next step will be to monitor all these parameters in the brain and to evaluate the effects of HZ in microglial cells and Nϕ to define the participation of this parasite metabolite on the development of CM.

As to the molecular mechanisms underlying the inflammatory response during CM, a recent study points to a role for PTK and NF-κB activity (853). Examination of the signaling events in the brain of mice suffering from fatal CM revealed increased PTK activity and tyrosine phosphorylation of proteins during CM pathology. Similarly, increased levels of NF-κB (p50 and p65 subunits) were detected in the nuclear extracts of the cerebral cortex and cerebellum. According to our data, HZ reduces Mϕ PTP activity (see Figure 10) and enhances nuclear translocation of both p50 and p65 NF-κB subunits in Mϕ (Figure 17) (776), two events that were associated with increased chemokine expression in these cells. Thus, it is likely that by triggering these and, probably, other intracellular events in one or more cellular populations in the brain HZ participates in proinflammatory mediator production in this organ. Further investigation will bring light on this important matter.

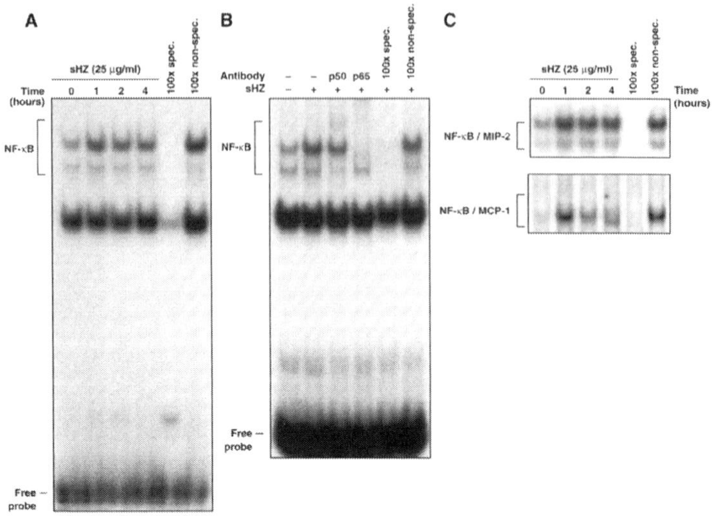

Figure 17. sHZ leads to NF-κB nuclear translocation and binding to the murine MIP-2 and MCP-1 promoters. (A) Nuclear extracts from Mϕ either left untreated or stimulated with sHZ for different time periods (0-4 h) were incubated with a [γ-^{32}P]-labeled NF-κB probe and were subjected to EMSA. (B) For supershift assays, nuclear extracts from cells stimulated with sHZ (1 h) were incubated or not with specific Abs against the p50 and p65 NF-κB isoforms for 1 h before EMSA. (C) EMSA analyses were performed as described in (A) but this time nuclear proteins were incubated with one of the NF-κB probes specific for the murine MIP-2 (*upper panel*) and MCP-1 (*lower panel*) promoters. Reprinted with permission, from *The Journal of Immunology*. Vol 174:475-484. Jaramillo, M., M. Godbout, and M. Olivier. Hemozoin induces macrophage chemokine expression through oxidative stress-dependent and -independent mechanisms. ***Copyright 2005. The American Association of Immunologists, Inc***. Reference (776).

Beyond its direct effects, it is possible that HZ exerts its actions through the induction of other molecules that might themselves contribute to CM, as seems to be the case of HETEs. In presence of HZ, AA was oxidized to HETEs, mainly 5- 12- and 15-HETE (409). These compounds are important vasoactive mediators (410), which can alter endothelial cell permeability (411), stimulate RBC adherence to capillary endothelium (412) and induce chemotaxis and chemokinesis (413). Moreover, large amounts of AA are released from cell lipids in response to brain injury (854) and its oxidation products appear to contribute to cerebral edema (411). Further supporting a role for HETEs in malaria pathology, it was recently reported that both trophozoite-iRBC and isolated HZ contained large amounts of hydroxy fatty acids, including the hydroxy-AA isomers 9-, 11- 12- and 15-HETE (201). Thus, it is conceivable that HZ-mediated AA oxidation to HETEs might be related to some of the pathophysiology associated with CM (*e.g.*, enhanced vascular permeability, edema, sequestration of iRBC).

In addition to a role for inflammatory responses, metabolic disturbances also contribute to severe malaria. Patients may suffer systemic, single- or multi-organ involvement, including acidosis, acute respiratory distress, coagulopathy, shock, hypoglycemia, renal dysfunction or failure, pulmonary edema, and severe hemoglobinuria (134). Acidosis and hypoglycemia have been associated with the worst outcomes in individuals with severe malaria infections (134, 151). That parasitic components might play an active role in such metabolic changes is supported by two studies performed by Schofield and colleagues (60, 229). First, *P. falciparum* GPI was shown to regulate glucose metabolism by adipocytes and to induce profound hypoglycemia *in vivo* (60). More recently, it was demonstrated that immunization against synthetic GPI substantially protects against cerebral syndrome, pulmonary edema, acidosis and fatality caused by *P. berghei* ANKA infection (229). These data show that severe disease parameters in the host are dependent on the actions of parasitic components, and they open the avenue for future investigation on the contribution of HZ to the metabolic disturbances associated to severe malaria.

It is well established that B cells and Abs are key players in immunity to malaria (84, 86, 90). Therefore, a key question that remains to be solved is related to the clinical protection of immunization against HZ. However, different lines of evidence indicate that this is a promising area of research. First, the production of significant amounts of anti-HZ IgM Abs was reported in *P. falciparum* malaria patients (205). Second, these Abs exhibited anti-inflammatory properties *in vitro* since they blocked HZ-inducible TNF-α and IL-β production by PBMC (205). These data suggest that anti-HZ Abs raised during infection may have a significant role in determining the severity of the disease by controlling the excessive production of cytokines. It will be of paramount importance to evaluate whether circulating levels of anti-HZ Abs correlate with protective immunity in endemic populations, as observed in the case of anti-GPI Abs (227). Despite these findings, two subsequent reports in which the correlation between anti-GPI Abs and malaria "tolerance" (*i.e.*, controlled parasitemias without symptoms) was analyzed, revealed that anti-GPI Abs, are unlikely to be the sole mediators against disease progression, mainly in children younger than five years (855, 856), further supporting the hypothesis that Abs against other parasitic components, such as HZ, might also confer some protection. Third, evaluation of the effect of HZ on DNA vaccine-induced Ab responses revealed that HZ enhanced Ab isotype switching during primary immunization, warranting further investigation of HZ as a potential immunomodulator (69). Fourth, our data demonstrated a causal relationship between HZ and inflammatory mediator *in vivo* (775); however, much needs to be done to validate HZ as a target antigen. Important points to be addressed include: (i) to define the contribution of HZ to local inflammation (*e.g.,* brain, placenta) and metabolic disturbances associated to malaria pathology; (ii) to evaluate the intensity, prevalence and specificity of the serological anti-HZ Ab response in mice immunized against purified HZ; (iii) to assess the protective effects of immunization against HZ in *Plasmodium*-infected mice (*e.g.,* acidosis, hypoglycemia, cerebral syndrome); (iv) to test the anti-inflammatory properties of anti-HZ Abs both *in vitro* and *in vivo*. Altogether, these studies will provide important information to define whether HZ is causally involved in malaria pathogenesis and whether vaccination against this target confers clinical protection. Even though this may provide an initial model for the development of an anti-disease vaccination against malaria, whether HZ is a target of clinical immunity in humans also remains elusive.

The design of a single anti-parasitic vaccine is extremely challenging. Among the numerous constraints that limit its effectiveness are the lack of sterile immunity and the complexity of the parasite life cycle, along with the existence of stage-specific parasite antigens, parasite protein polymorphisms, and infections with heterologous *Plasmodium* species (34). In this context, the design of anti-disease vaccine that may reduce or suppress the pathogenecity of the disease without requiring eradication of the parasite, would protect the groups at higher risk of dying from severe malaria; that is, children, pregnant women and migrants who live in the sub-Saharan Africa, where 90% of deaths (1 to 3 million per year) associated with malaria occur. Another the important advantage of designing an anti-disease vaccine is its potential to simultaneously block the actions of numerous inflammatory mediators, in contrast to antimalarial therapy based on cytokine-specific Abs and/or inhibitors. In fact, such approaches have experienced limited success, which is probably related to the fact that many cytokines have similar and overlapping functions (95). In addition, when evaluating an anti-GPI versus an anti-HZ vaccine, it will be important to take into account that because of its mode of synthesis, it is likely that the large-scale production of an anti-HZ vaccine results in much lower costs than one based on the synthetic preparation of GPI. Detailed knowledge of the structure-activity and structure-immunogenicity relationships of HZ may offer insight into the design of HZ-based therapeutics or a vaccine for malaria disease.

In conclusion, our studies demonstrate that HZ is a potent proinflammatory agent *in vitro* and *in vivo*, and suggest that this molecule, which is released in large amounts from iRBC, may contribute to mediate both beneficial and detrimental effects on the infected host. Because erythrocytic schizogeny and subsequent HZ release occur in the microvasculature of multiple organs (*e.g.,* liver, placenta, brain) where iRBC are sequestered, and both HZ and leukocyte accumulation have been reported in these locations, it is possible that the HZ-inducible immunological responses would be directed against the sites of maximal parasite sequestration. Therefore, by triggering specific intracellular signals leading to immune cell activation, HZ might function as a “danger signal” to alert the immune system against pathogen invasion. However, through an excessive inflammatory response, HZ is also likely to contribute to the pathogenesis of severe malaria. Although a causal link between HZ and the clinical manifestations of malaria remain to be established, our data -along with previous evidence from both human and experimental malaria- warrants further investigation of HZ as an immunomodulator. A better understanding of the antigenic properties of HZ as well as the analysis of (i) the local and temporal effects of HZ *in vivo*, (ii) the cellular sources and (iii) the signal transduction mechanisms through which this parasite metabolite controls the immune response, will be helpful in the design of future immune treatments against malaria to prevent millions of deaths.

REFERENCES

1. Vivier, E., and I. Desportes. 1989. Apicomplexa. In *Hadbook of Protocista.* L. Margulis, J. O. Corliss, M. Melkonia, and D. J. Chapman, eds. Jones and Barlett, Boston.
2. Ayala, F. J., A. A. Escalante, A. A. Lal, and S. M. Rich. 1998. Evolutionary Relationships of Human Malaria Parasites. In *Malaria: Parasite Biology, Pathogenesis, and Protection.* I. W. Sherman, ed. ASM Press, Washington D. C.
3. Levine, N. D. 1988. *The Protozoan Phylum Apicomplexa.* CRC Press. Boca Raton.
4. Snewin, V. A., S. Longacre, and P. H. David. 1991. Plasmodium vivax: older and wiser? *Res Immunol* 142:631-636.
5. Waters, A. P., D. G. Higgins, and T. F. McCutchan. 1991. Plasmodium falciparum appears to have arisen as a result of lateral transfer between avian and human hosts. *Proc Natl Acad Sci U S A* 88:3140-3144.
6. Ayala, F. J., and W. M. Fitch. 1992. Phylogeny of plasmodium falciparum. *Parasitol Today* 8:74-75.
7. Torii, M., and M. Aikawa. 1998. Ultrastructure of Asexual Stages. In *Malaria: Parasite Biology, Pathogenesis, and Protection.* I. W. Sherman, ed. ASM Press, Washington D. C.
8. Bannister, L., and G. Mitchell. 2003. The ins, outs and roundabouts of malaria. *Trends Parasitol* 19:209-213.
9. Fujioka, H., and M. Aikawa. 2002. Structure and life cycle. *Chem Immunol* 80:1-26.
10. Szarfman, A., J. A. Lyon, D. Walliker, I. Quakyi, R. J. Howard, S. Sun, W. R. Ballou, K. Esser, W. T. London, R. A. Wirtz, and et al. 1988. Mature liver stages of cloned Plasmodium falciparum share epitopes with proteins from sporozoites and asexual blood stages. *Parasite Immunol* 10:339-351.
11. Krotoski, W. A., W. E. Collins, R. S. Bray, P. C. Garnham, F. B. Cogswell, R. W. Gwadz, R. Killick-Kendrick, R. Wolf, R. Sinden, L. C. Koontz, and P. S. Stanfill. 1982. Demonstration of hypnozoites in sporozoite-transmitted Plasmodium vivax infection. *Am J Trop Med Hyg* 31:1291-1293.
12. Roberts, L., and J. Janovy. 1999 *Foundations of Parasitology.* McGraw Hill, New York.
13. Chitnis, C. E., and M. J. Blackman. 2000. Host cell invasion by malaria parasites. *Parasitol Today* 16:411-415.
14. Langreth, S. G., J. B. Jensen, R. T. Reese, and W. Trager. 1978. Fine structure of human malaria in vitro. *J Protozool* 25:443-452.
15. Slater, A. F., and A. Cerami. 1992. Inhibition by chloroquine of a novel haem polymerase enzyme activity in malaria trophozoites. *Nature* 355:167-169.
16. Berendt, A. R., D. J. Ferguson, and C. I. Newbold. 1990. Sequestration in Plasmodium falciparum malaria: sticky cells and sticky problems. *Parasitol Today* 6:247-254.
17. Adams, S., H. Brown, and G. Turner. 2002. Breaking down the blood-brain barrier: signaling a path to cerebral malaria? *Trends Parasitol* 18:360-366.
18. Sachs, J., and P. Malaney. 2002. The economic and social burden of malaria. *Nature* 415:680-685.
19. Breman, J. G. 2001. The ears of the hippopotamus: manifestations, determinants, and estimates of the malaria burden. *Am J Trop Med Hyg* 64:1-11.
20. Greenwood, B., and T. Mutabingwa. 2002. Malaria in 2002. *Nature* 415:670-672.
21. Ridley, R. G. 2002. Medical need, scientific opportunity and the drive for antimalarial drugs. *Nature* 415:686-693.
22. Roberts, D. R., S. Manguin, and J. Mouchet. 2000. DDT house spraying and re-emerging malaria. *Lancet* 356:330-332.
23. Collins, F. H., L. Kamau, H. A. Ranson, and J. M. Vulule. 2000. Molecular entomology and prospects for malaria control. *Bull World Health Organ* 78:1412-1423.
24. Kidane, G., and R. H. Morrow. 2000. Teaching mothers to provide home treatment of malaria in Tigray, Ethiopia: a randomised trial. *Lancet* 356:550-555.
25. Bloland, P. B., and M. Ettling. 1999. Making malaria-treatment policy in the face of drug resistance. *Ann Trop Med Parasitol* 93:5-23.
26. Hoffman, S. L., C. N. Oster, C. V. Plowe, G. R. Woollett, J. C. Beier, J. D. Chulay, R. A. Wirtz, M. R. Hollingdale, and M. Mugambi. 1987. Naturally acquired antibodies to sporozoites do not prevent malaria: vaccine development implications. *Science* 237:639-642.
27. Goodman, C. A., P. G. Coleman, and A. J. Mills. 1999. Cost-effectiveness of malaria control in sub-Saharan Africa. *Lancet* 354:378-385.
28. White, N. 1999. Antimalarial drug resistance and combination chemotherapy. *Philos Trans R Soc Lond B Biol Sci* 354:739-749.
29. Wellems, T. E., and C. V. Plowe. 2001. Chloroquine-resistant malaria. *J Infect Dis* 184:770-776.
30. Takechi, M., M. Matsuo, C. Ziba, A. MacHeso, D. Butao, I. L. Zungu, I. Chakanika, and M. D. Bustos. 2001. Therapeutic efficacy of sulphadoxine/pyrimethamine and susceptibility in vitro of P. falciparum isolates to sulphadoxine-pyremethamine and other antimalarial drugs in Malawian children. *Trop Med Int Health* 6:429-434.

31. Guerin, P. J., P. Olliaro, F. Nosten, P. Druilhe, R. Laxminarayan, F. Binka, W. L. Kilama, N. Ford, and N. J. White. 2002. Malaria: current status of control, diagnosis, treatment, and a proposed agenda for research and development. *Lancet Infect Dis* 2:564-573.
32. Adjuik, M., P. Agnamey, A. Babiker, S. Borrmann, P. Brasseur, M. Cisse, F. Cobelens, S. Diallo, J. F. Faucher, P. Garner, S. Gikunda, P. G. Kremsner, S. Krishna, B. Lell, M. Loolpapit, P. B. Matsiegui, M. A. Missinou, J. Mwanza, F. Ntoumi, P. Olliaro, P. Osimbo, P. Rezbach, E. Some, and W. R. Taylor. 2002. Amodiaquine-artesunate versus amodiaquine for uncomplicated Plasmodium falciparum malaria in African children: a randomised, multicentre trial. *Lancet* 359:1365-1372.
33. Hoffman, S. L., G. M. Subramanian, F. H. Collins, and J. C. Venter. 2002. Plasmodium, human and Anopheles genomics and malaria. *Nature* 415:702-709.
34. Richie, T. L., and A. Saul. 2002. Progress and challenges for malaria vaccines. *Nature* 415:694-701.
35. Saul, A. 1999. The role of variant surface antigens on malaria-infected red blood cells. *Parasitol Today* 15:455-457.
36. Clyde, D. F. 1990. Immunity to falciparum and vivax malaria induced by irradiated sporozoites: a review of the University of Maryland studies, 1971-75. *Bull World Health Organ* 68 Suppl:9-12.
37. Kumar, S., J. E. Epstein, T. L. Richie, F. K. Nkrumah, L. Soisson, D. J. Carucci, and S. L. Hoffman. 2002. A multilateral effort to develop DNA vaccines against falciparum malaria. *Trends Parasitol* 18:129-135.
38. Doolan, D. L., and S. L. Hoffman. 2000. The complexity of protective immunity against liver-stage malaria. *J Immunol* 165:1453-1462.
39. Wipasa, J., S. Elliott, H. Xu, and M. F. Good. 2002. Immunity to asexual blood stage malaria and vaccine approaches. *Immunol Cell Biol* 80:401-414.
40. Jones, T. R., N. Obaldia, 3rd, R. A. Gramzinski, Y. Charoenvit, N. Kolodny, S. Kitov, H. L. Davis, A. M. Krieg, and S. L. Hoffman. 1999. Synthetic oligodeoxynucleotides containing CpG motifs enhance immunogenicity of a peptide malaria vaccine in Aotus monkeys. *Vaccine* 17:3065-3071.
41. Baird, J. K. 1998. Age-dependent characteristics of protection v. susceptibility to Plasmodium falciparum. *Ann Trop Med Parasitol* 92:367-390.
42. Gupta, S., R. W. Snow, C. A. Donnelly, K. Marsh, and C. Newbold. 1999. Immunity to non-cerebral severe malaria is acquired after one or two infections. *Nat Med* 5:340-343.
43. Andrysiak, P. M., W. E. Collins, and G. H. Campbell. 1986. Stage-specific and species-specific antigens of Plasmodium vivax and Plasmodium ovale defined by monoclonal antibodies. *Infect Immun* 54:609-612.
44. Nussenzweig, V., and R. S. Nussenzweig. 1989. Rationale for the development of an engineered sporozoite malaria vaccine. *Adv Immunol* 45:283-334.
45. Nardin, E. H., R. S. Nussenzweig, I. A. McGregor, and J. H. Bryan. 1979. Antibodies to sporozoites: their frequent occurrence in individuals living in an area of hyperendemic malaria. *Science* 206:597-599.
46. Potocnjak, P., N. Yoshida, R. S. Nussenzweig, and V. Nussenzweig. 1980. Monovalent fragments (Fab) of monoclonal antibodies to a sporozoite surface antigen (Pb44) protect mice against malarial infection. *J Exp Med* 151:1504-1513.
47. Schofield, L., J. Villaquiran, A. Ferreira, H. Schellekens, R. Nussenzweig, and V. Nussenzweig. 1987. Gamma interferon, CD8+ T cells and antibodies required for immunity to malaria sporozoites. *Nature* 330:664-666.
48. Tsuji, M., P. Romero, R. S. Nussenzweig, and F. Zavala. 1990. CD4+ cytolytic T cell clone confers protection against murine malaria. *J Exp Med* 172:1353-1357.
49. Sinnis, P., and E. Nardin. 2002. Sporozoite antigens: biology and immunology of the circumsporozoite protein and thrombospondin-related anonymous protein. *Chem Immunol* 80:70-96.
50. Druilhe, P., and J. L. Perignon. 1994. Mechanisms of defense against P. falciparum asexual blood stages in humans. *Immunol Lett* 41:115-120.
51. Ferrante, A., L. Kumaratilake, C. M. Rzepczyk, and J. M. Dayer. 1990. Killing of Plasmodium falciparum by cytokine activated effector cells (neutrophils and macrophages). *Immunol Lett* 25:179-187.
52. Phillips, S. 1994. Effector mechanisms against asexual erythrocytic stages of Plasmodium. *Immunol Lett* 41:109-114.
53. Sheagren, J. N., J. E. Tobie, L. M. Fox, and S. M. Wolff. 1970. Reticuloendothelial system phagocytic function in naturally acquired human malaria. *J Lab Clin Med* 75:481-487.
54. Jones, K. R., B. J. Cottrell, G. A. Targett, and J. H. Playfair. 1989. Killing of Plasmodium falciparum by human monocyte-derived macrophages. *Parasite Immunol* 11:585-592.
55. Kumaratilake, L. M., and A. Ferrante. 1992. IL-4 inhibits macrophage-mediated killing of Plasmodium falciparum in vitro. A possible parasite-immune evasion mechanism. *J Immunol* 149:194-199.
56. Ockenhouse, C. F., and H. L. Shear. 1984. Oxidative killing of the intraerythrocytic malaria parasite Plasmodium yoelii by activated macrophages. *J Immunol* 132:424-431.
57. Wozencraft, A. O., S. L. Croft, and G. Sayers. 1985. Oxygen radical release by adherent cell populations during the initial stages of a lethal rodent malarial infection. *Immunology* 56:523-531.
58. Dubey, M. L., S. K. Rai, N. K. Ganguly, A. Kalra, S. C. Varma, and R. C. Mahajan. 1991. Generation of reactive oxygen species by blood monocytes in human Plasmodium falciparum and P. vivax infections. *APMIS* 99:210-212.

59. Kremsner, P. G., A. Nussler, S. Neifer, M. F. Chaves, U. Bienzle, G. Senaldi, and G. E. Grau. 1993. Malaria antigen and cytokine-induced production of reactive nitrogen intermediates by murine macrophages: no relevance to the development of experimental cerebral malaria. *Immunology* 78:286-290.

60. Schofield, L., and F. Hackett. 1993. Signal transduction in host cells by a glycosylphosphatidylinositol toxin of malaria parasites. *J Exp Med* 177:145-153.

61. Sherry, B. A., G. Alava, K. J. Tracey, J. Martiney, A. Cerami, and A. F. Slater. 1995. Malaria-specific metabolite hemozoin mediates the release of several potent endogenous pyrogens (TNF, MIP-1 alpha, and MIP-1 beta) in vitro, and altered thermoregulation in vivo. *J Inflamm* 45:85-96.

62. Picot, S., F. Peyron, P. Deloron, C. Boudin, B. Chumpitazi, G. Barbe, J. P. Vuillez, A. Donadille, and P. Ambroise-Thomas. 1993. Ring-infected erythrocyte surface antigen (Pf/155RESA) induces tumour necrosis factor-alpha production. *Clin Exp Immunol* 93:184-188.

63. Taverne, J., C. A. Bate, D. Kwiatkowski, P. H. Jakobsen, and J. H. Playfair. 1990. Two soluble antigens of Plasmodium falciparum induce tumor necrosis factor release from macrophages. *Infect Immun* 58:2923-2928.

64. Jakobsen, P. H., R. Moon, R. G. Ridley, C. A. Bate, J. Taverne, M. B. Hansen, B. Takacs, J. H. Playfair, and J. S. McBride. 1993. Tumour necrosis factor and interleukin-6 production induced by components associated with merozoite proteins of Plasmodium falciparum. *Parasite Immunol* 15:229-237.

65. Pichyangkul, S., P. Saengkrai, and H. K. Webster. 1994. Plasmodium falciparum pigment induces monocytes to release high levels of tumor necrosis factor-alpha and interleukin-1 beta. *Am J Trop Med Hyg* 51:430-435.

66. Tachado, S. D., P. Gerold, M. J. McConville, T. Baldwin, D. Quilici, R. T. Schwarz, and L. Schofield. 1996. Glycosylphosphatidylinositol toxin of Plasmodium induces nitric oxide synthase expression in macrophages and vascular endothelial cells by a protein tyrosine kinase-dependent and protein kinase C-dependent signaling pathway. *J Immunol* 156:1897-1907.

67. Schwarzer, E., F. Turrini, D. Ulliers, G. Giribaldi, H. Ginsburg, and P. Arese. 1992. Impairment of macrophage functions after ingestion of Plasmodium falciparum-infected erythrocytes or isolated malarial pigment. *J Exp Med* 176:1033-1041.

68. Seixas, E., C. Cross, S. Quin, and J. Langhorne. 2001. Direct activation of dendritic cells by the malaria parasite, Plasmodium chabaudi chabaudi. *Eur J Immunol* 31:2970-2978.

69. Coban, C., K. J. Ishii, D. J. Sullivan, and N. Kumar. 2002. Purified malaria pigment (hemozoin) enhances dendritic cell maturation and modulates the isotype of antibodies induced by a DNA vaccine. *Infect Immun* 70:3939-3943.

70. Kumaratilake, L. M., A. Ferrante, J. S. Kumaratilake, and A. C. Allison. 1994. Extraction of intraerythrocytic malarial parasites by phagocytic cells. *Parasitol Today* 10:193-196.

71. Brown, J., and M. E. Smalley. 1981. Inhibition of the in vitro growth of Plasmodium falciparum by human polymorphonuclear neutrophil leucocytes. *Clin Exp Immunol* 46:106-109.

72. Orago, A. S., and C. A. Facer. 1991. Cytotoxicity of human natural killer (NK) cell subsets for Plasmodium falciparum erythrocytic schizonts: stimulation by cytokines and inhibition by neomycin. *Clin Exp Immunol* 86:22-29.

73. Mohan, K., P. Moulin, and M. M. Stevenson. 1997. Natural killer cell cytokine production, not cytotoxicity, contributes to resistance against blood-stage Plasmodium chabaudi AS infection. *J Immunol* 159:4990-4998.

74. Aderem, A., and R. J. Ulevitch. 2000. Toll-like receptors in the induction of the innate immune response. *Nature* 406:782-787.

75. Campos, M. A., I. C. Almeida, O. Takeuchi, S. Akira, E. P. Valente, D. O. Procopio, L. R. Travassos, J. A. Smith, D. T. Golenbock, and R. T. Gazzinelli. 2001. Activation of Toll-like receptor-2 by glycosylphosphatidylinositol anchors from a protozoan parasite. *J Immunol* 167:416-423.

76. Scanga, C. A., J. Aliberti, D. Jankovic, F. Tilloy, S. Bennouna, E. Y. Denkers, R. Medzhitov, and A. Sher. 2002. Cutting edge: MyD88 is required for resistance to Toxoplasma gondii infection and regulates parasite-induced IL-12 production by dendritic cells. *J Immunol* 168:5997-6001.

77. Kropf, P., N. Freudenberg, C. Kalis, M. Modolell, S. Herath, C. Galanos, M. Freudenberg, and I. Muller. 2004. Infection of C57BL/10ScCr and C57BL/10ScNCr mice with Leishmania major reveals a role for Toll-like receptor 4 in the control of parasite replication. *J Leukoc Biol* 76:48-57.

78. Adachi, K., H. Tsutsui, S. Kashiwamura, E. Seki, H. Nakano, O. Takeuchi, K. Takeda, K. Okumura, L. Van Kaer, H. Okamura, S. Akira, and K. Nakanishi. 2001. Plasmodium berghei infection in mice induces liver injury by an IL-12- and toll-like receptor/myeloid differentiation factor 88-dependent mechanism. *J Immunol* 167:5928-5934.

79. Pichyangkul, S., K. Yongvanitchit, U. Kum-arb, H. Hemmi, S. Akira, A. M. Krieg, D. G. Heppner, V. A. Stewart, H. Hasegawa, S. Looareesuwan, G. D. Shanks, and R. S. Miller. 2004. Malaria blood stage parasites activate human plasmacytoid dendritic cells and murine dendritic cells through a Toll-like receptor 9-dependent pathway. *J Immunol* 172:4926-4933.

80. Perlmann, P., and M. Troye-Blomberg. 2002. Malaria and the immune system in humans. *Chem Immunol* 80:229-242.

81. Chemtai, A. K., C. Hamers-Casterman, R. Hamers, and P. De Baetselier. 1984. T cell-mediated immunity in murine malaria. II. Induction of protective immunity to P. chabaudi by antigen-fed macrophages and antigen-educated lymphocytes. *Parasite Immunol* 6:469-480.
82. Bruna-Romero, O., and A. Rodriguez. 2001. Dendritic cells can initiate protective immune responses against malaria. *Infect Immun* 69:5173-5176.
83. Van der Heyde, H. C., D. Huszar, C. Woodhouse, D. D. Manning, and W. P. Weidanz. 1994. The resolution of acute malaria in a definitive model of B cell deficiency, the JHD mouse. *J Immunol* 152:4557-4562.
84. Langhorne, J., C. Cross, E. Seixas, C. Li, and T. von der Weid. 1998. A role for B cells in the development of T cell helper function in a malaria infection in mice. *Proc Natl Acad Sci U S A* 95:1730-1734.
85. Constant, S. L., and K. Bottomly. 1997. Induction of Th1 and Th2 CD4+ T cell responses: the alternative approaches. *Annu Rev Immunol* 15:297-322.
86. Taylor-Robinson, A. W., and R. S. Phillips. 1994. B cells are required for the switch from Th1- to Th2-regulated immune responses to Plasmodium chabaudi chabaudi infection. *Infect Immun* 62:2490-2498.
87. Langhorne, J., S. J. Meding, K. Eichmann, and S. S. Gillard. 1989. The response of CD4+ T cells to Plasmodium chabaudi chabaudi. *Immunol Rev* 112:71-94.
88. Bull, P. C., B. S. Lowe, M. Kortok, C. S. Molyneux, C. I. Newbold, and K. Marsh. 1998. Parasite antigens on the infected red cell surface are targets for naturally acquired immunity to malaria. *Nat Med* 4:358-360.
89. Bouharoun-Tayoun, H., P. Attanath, A. Sabchareon, T. Chongsuphajaisiddhi, and P. Druilhe. 1990. Antibodies that protect humans against Plasmodium falciparum blood stages do not on their own inhibit parasite growth and invasion in vitro, but act in cooperation with monocytes. *J Exp Med* 172:1633-1641.
90. Spencer Valero, L. M., S. A. Ogun, S. L. Fleck, I. T. Ling, T. J. Scott-Finnigan, M. J. Blackman, and A. A. Holder. 1998. Passive immunization with antibodies against three distinct epitopes on Plasmodium yoelii merozoite surface protein 1 suppresses parasitemia. *Infect Immun* 66:3925-3930.
91. English, M., and C. R. Newton. 2002. Malaria: pathogenicity and disease. *Chem Immunol* 80:50-69.
92. Clark, I. A., and N. H. Hunt. 1983. Evidence for reactive oxygen intermediates causing hemolysis and parasite death in malaria. *Infect Immun* 39:1-6.
93. Greve, B., P. G. Kremsner, B. Lell, D. Luckner, and D. Schmid. 2000. Malarial anaemia in African children associated with high oxygen-radical production. *Lancet* 355:40-41.
94. Das, B. S., S. Mohanty, S. K. Mishra, J. K. Patnaik, S. K. Satpathy, D. Mohanty, and T. K. Bose. 1991. Increased cerebrospinal fluid protein and lipid peroxidation products in patients with cerebral malaria. *Trans R Soc Trop Med Hyg* 85:733-734.
95. Urquhart, A. D. 1994. Putative pathophysiological interactions of cytokines and phagocytic cells in severe human falciparum malaria. *Clin Infect Dis* 19:117-131.
96. Kern, P., C. J. Hemmer, J. Van Damme, H. J. Gruss, and M. Dietrich. 1989. Elevated tumor necrosis factor alpha and interleukin-6 serum levels as markers for complicated Plasmodium falciparum malaria. *Am J Med* 87:139-143.
97. Mshana, R. N., J. Boulandi, N. M. Mshana, J. Mayombo, and G. Mendome. 1991. Cytokines in the pathogenesis of malaria: levels of IL-I beta, IL-4, IL-6, TNF-alpha and IFN-gamma in plasma of healthy individuals and malaria patients in a holoendemic area. *J Clin Lab Immunol* 34:131-139.
98. Day, N. P., T. T. Hien, T. Schollaardt, P. P. Loc, L. V. Chuong, T. T. Chau, N. T. Mai, N. H. Phu, D. X. Sinh, N. J. White, and M. Ho. 1999. The prognostic and pathophysiologic role of pro- and antiinflammatory cytokines in severe malaria. *J Infect Dis* 180:1288-1297.
99. Burgmann, H., U. Hollenstein, C. Wenisch, F. Thalhammer, S. Looareesuwan, and W. Graninger. 1995. Serum concentrations of MIP-1 alpha and interleukin-8 in patients suffering from acute Plasmodium falciparum malaria. *Clin Immunol Immunopathol* 76:32-36.
100. Friedland, J. S., M. Ho, D. G. Remick, D. Bunnag, N. J. White, and G. E. Griffin. 1993. Interleukin-8 and Plasmodium falciparum malaria in Thailand. *Trans R Soc Trop Med Hyg* 87:54-55.
101. Bordmann, G., G. Burmeister, S. Saladin, H. Urassa, S. Mwankyusye, N. Weiss, and M. Tanner. 1997. MRP 8/14 as marker for Plasmodium falciparum-induced malaria episodes in individuals in a holoendemic area. *Clin Diagn Lab Immunol* 4:435-439.
102. Kurtzhals, J. A., V. Adabayeri, B. Q. Goka, B. D. Akanmori, J. O. Oliver-Commey, F. K. Nkrumah, C. Behr, and L. Hviid. 1998. Low plasma concentrations of interleukin 10 in severe malarial anaemia compared with cerebral and uncomplicated malaria. *Lancet* 351:1768-1772.
103. Malaguarnera, L., and S. Musumeci. 2002. The immune response to Plasmodium falciparum malaria. *Lancet Infect Dis* 2:472-478.
104. Urban, B. C., and D. J. Roberts. 2002. Malaria, monocytes, macrophages and myeloid dendritic cells: sticking of infected erythrocytes switches off host cells. *Curr Opin Immunol* 14:458-465.
105. Plebanski, M., and A. V. Hill. 2000. The immunology of malaria infection. *Curr Opin Immunol* 12:437-441.
106. Ho, M., and H. K. Webster. 1989. Immunology of human malaria. A cellular perspective. *Parasite Immunol* 11:105-116.
107. Mabey, D. C., A. Brown, and B. M. Greenwood. 1987. Plasmodium falciparum malaria and Salmonella infections in Gambian children. *J Infect Dis* 155:1319-1321.

108. Cook, I. F. 1985. Herpes zoster in children following malaria. *J Trop Med Hyg* 88:261-264.
109. Wedderburn, N., and B. N. Dracott. 1977. The immune reponse to type III pneumococcal polysaccharide in mice with malaria. *Clin Exp Immunol* 28:130-137.
110. Weidanz, W. P. 1982. Malaria and alterations in immune reactivity. *Br Med Bull* 38:167-172.
111. Lelchuk, R., G. Rose, and J. H. Playfair. 1984. Changes in the capacity of macrophages and T cells to produce interleukins during murine malaria infection. *Cell Immunol* 84:253-263.
112. Rockett, K. A., M. M. Awburn, E. J. Rockett, W. B. Cowden, and I. A. Clark. 1994. Possible role of nitric oxide in malarial immunosuppression. *Parasite Immunol* 16:243-249.
113. Stevenson, M. M., E. Ghadirian, N. C. Phillips, D. Rae, and J. E. Podoba. 1989. Role of mononuclear phagocytes in elimination of Plasmodium chabaudi AS infection. *Parasite Immunol* 11:529-544.
114. Ward, K. N., M. J. Warrell, J. Rhodes, S. Looareesuwan, and N. J. White. 1984. Altered expression of human monocyte Fc receptors in Plasmodium falciparum malaria. *Infect Immun* 44:623-626.
115. Shear, H. L. 1988. Variation in expression of antibody-dependent cell-mediated cytotoxicity in rodents with malaria. *Infect Immun* 56:3007-3010.
116. Clark, I. A., J. L. Virelizier, E. A. Carswell, and P. R. Wood. 1981. Possible importance of macrophage-derived mediators in acute malaria. *Infect Immun* 32:1058-1066.
117. Ho, M., N. J. White, S. Looareesuwan, Y. Wattanagoon, S. H. Lee, M. J. Walport, D. Bunnag, and T. Harinasuta. 1990. Splenic Fc receptor function in host defense and anemia in acute Plasmodium falciparum malaria. *J Infect Dis* 161:555-561.
118. Ho, M., H. K. Webster, S. Looareesuwan, W. Supanaranond, R. E. Phillips, P. Chanthavanich, and D. A. Warrell. 1986. Antigen-specific immunosuppression in human malaria due to Plasmodium falciparum. *J Infect Dis* 153:763-771.
119. Morakote, N., and D. E. Justus. 1988. Immunosuppression in malaria: effect of hemozoin produced by Plasmodium berghei and Plasmodium falciparum. *Int Arch Allergy Appl Immunol* 86:28-34.
120. Scorza, T., S. Magez, L. Brys, and P. De Baetselier. 1999. Hemozoin is a key factor in the induction of malaria-associated immunosuppression. *Parasite Immunol* 21:545-554.
121. Taramelli, D., N. Basilico, E. Pagani, R. Grande, D. Monti, M. Ghione, and P. Olliaro. 1995. The heme moiety of malaria pigment (beta-hematin) mediates the inhibition of nitric oxide and tumor necrosis factor-alpha production by lipopolysaccharide-stimulated macrophages. *Exp Parasitol* 81:501-511.
122. Schwarzer, E., M. Alessio, D. Ulliers, and P. Arese. 1998. Phagocytosis of the malarial pigment, hemozoin, impairs expression of major histocompatibility complex class II antigen, CD54, and CD11c in human monocytes. *Infect Immun* 66:1601-1606.
123. Urban, B. C., D. J. Ferguson, A. Pain, N. Willcox, M. Plebanski, J. M. Austyn, and D. J. Roberts. 1999. Plasmodium falciparum-infected erythrocytes modulate the maturation of dendritic cells. *Nature* 400:73-77.
124. Urban, B. C., N. Willcox, and D. J. Roberts. 2001. A role for CD36 in the regulation of dendritic cell function. *Proc Natl Acad Sci U S A* 98:8750-8755.
125. Miller, L. H., D. I. Baruch, K. Marsh, and O. K. Doumbo. 2002. The pathogenic basis of malaria. *Nature* 415:673-679.
126. Beeson, J. G., and G. V. Brown. 2002. Pathogenesis of Plasmodium falciparum malaria: the roles of parasite adhesion and antigenic variation. *Cell Mol Life Sci* 59:258-271.
127. Greenwood, B. M., A. K. Bradley, A. M. Greenwood, P. Byass, K. Jammeh, K. Marsh, S. Tulloch, F. S. Oldfield, and R. Hayes. 1987. Mortality and morbidity from malaria among children in a rural area of The Gambia, West Africa. *Trans R Soc Trop Med Hyg* 81:478-486.
128. Kwiatkowski, D. 1989. Febrile temperatures can synchronize the growth of Plasmodium falciparum in vitro. *J Exp Med* 169:357-361.
129. Brown, W. H. 1912. Malarial pigment (hematin) as a factor in the production of the malarial paroxysm. *J. Exp. Med.* 15:579.
130. Kwiatkowski, D., J. G. Cannon, K. R. Manogue, A. Cerami, C. A. Dinarello, and B. M. Greenwood. 1989. Tumour necrosis factor production in Falciparum malaria and its association with schizont rupture. *Clin Exp Immunol* 77:361-366.
131. Karunaweera, N. D., G. E. Grau, P. Gamage, R. Carter, and K. N. Mendis. 1992. Dynamics of fever and serum levels of tumor necrosis factor are closely associated during clinical paroxysms in Plasmodium vivax malaria. *Proc Natl Acad Sci U S A* 89:3200-3203.
132. Kwiatkowski, D., M. E. Molyneux, S. Stephens, N. Curtis, N. Klein, P. Pointaire, M. Smit, R. Allan, D. R. Brewster, G. E. Grau, and et al. 1993. Anti-TNF therapy inhibits fever in cerebral malaria. *Q J Med* 86:91-98.
133. Kwiatkowski, D. 1995. Malarial toxins and the regulation of parasite density. *Parasitol Today* 11:206-212.
134. White, N. J., and M. Ho. 1992. The pathophysiology of malaria. *Adv Parasitol* 31:83-173.
135. Marsh, K., M. English, J. Crawley, and N. Peshu. 1996. The pathogenesis of severe malaria in African children. *Ann Trop Med Parasitol* 90:395-402.
136. Bondi, F. S. 1992. The incidence and outcome of neurological abnormalities in childhood cerebral malaria: a long-term follow-up of 62 survivors. *Trans R Soc Trop Med Hyg* 86:17-19.

137. Grau, G. E., H. Heremans, P. F. Piguet, P. Pointaire, P. H. Lambert, A. Billiau, and P. Vassalli. 1989. Monoclonal antibody against interferon gamma can prevent experimental cerebral malaria and its associated overproduction of tumor necrosis factor. *Proc Natl Acad Sci U S A* 86:5572-5574.

138. Kwiatkowski, D., A. V. Hill, I. Sambou, P. Twumasi, J. Castracane, K. R. Manogue, A. Cerami, D. R. Brewster, and B. M. Greenwood. 1990. TNF concentration in fatal cerebral, non-fatal cerebral, and uncomplicated Plasmodium falciparum malaria. *Lancet* 336:1201-1204.

139. Brown, H., G. Turner, S. Rogerson, M. Tembo, J. Mwenechanya, M. Molyneux, and T. Taylor. 1999. Cytokine expression in the brain in human cerebral malaria. *J Infect Dis* 180:1742-1746.

140. Berendt, A. R., A. McDowall, A. G. Craig, P. A. Bates, M. J. Sternberg, K. Marsh, C. I. Newbold, and N. Hogg. 1992. The binding site on ICAM-1 for Plasmodium falciparum-infected erythrocytes overlaps, but is distinct from, the LFA-1-binding site. *Cell* 68:71-81.

141. Clark, I. A., K. A. Rockett, and W. B. Cowden. 1991. Proposed link between cytokines, nitric oxide and human cerebral malaria. *Parasitol Today* 7:205-207.

142. Al Yaman, F. M., D. Mokela, B. Genton, K. A. Rockett, M. P. Alpers, and I. A. Clark. 1996. Association between serum levels of reactive nitrogen intermediates and coma in children with cerebral malaria in Papua New Guinea. *Trans R Soc Trop Med Hyg* 90:270-273.

143. Anstey, N. M., J. B. Weinberg, M. Y. Hassanali, E. D. Mwaikambo, D. Manyenga, M. A. Misukonis, D. R. Arnelle, D. Hollis, M. I. McDonald, and D. L. Granger. 1996. Nitric oxide in Tanzanian children with malaria: inverse relationship between malaria severity and nitric oxide production/nitric oxide synthase type 2 expression. *J Exp Med* 184:557-567.

144. Maneerat, Y., P. Viriyavejakul, B. Punpoowong, M. Jones, P. Wilairatana, E. Pongponratn, G. D. Turner, and R. Udomsangpetch. 2000. Inducible nitric oxide synthase expression is increased in the brain in fatal cerebral malaria. *Histopathology* 37:269-277.

145. Deininger, M. H., P. G. Kremsner, R. Meyermann, and H. Schluesener. 2002. Macrophages/microglial cells in patients with cerebral malaria. *Eur Cytokine Netw* 13:173-185.

146. Patnaik, J. K., B. S. Das, S. K. Mishra, S. Mohanty, S. K. Satpathy, and D. Mohanty. 1994. Vascular clogging, mononuclear cell margination, and enhanced vascular permeability in the pathogenesis of human cerebral malaria. *Am J Trop Med Hyg* 51:642-647.

147. Schluesener, H. J., P. G. Kremsner, and R. Meyermann. 1998. Widespread expression of MRP8 and MRP14 in human cerebral malaria by microglial cells. *Acta Neuropathol* 96:575-580.

148. Ryckman, C., K. Vandal, P. Rouleau, M. Talbot, and P. A. Tessier. 2003. Proinflammatory activities of S100: proteins S100A8, S100A9, and S100A8/A9 induce neutrophil chemotaxis and adhesion. *J Immunol* 170:3233-3242.

149. Maeno, Y., P. Perlmann, PerlmannH, Y. Kusuhara, K. Taniguchi, T. Nakabayashi, K. Win, S. Looareesuwan, and M. Aikawa. 2000. IgE deposition in brain microvessels and on parasitized erythrocytes from cerebral malaria patients. *Am J Trop Med Hyg* 63:128-132.

150. Clark, I. A., and K. A. Rockett. 1994. The cytokine theory of human cerebral malaria. *Parasitol Today* 10:410-412.

151. English, M., C. Waruiru, E. Amukoye, S. Murphy, J. Crawley, I. Mwangi, N. Peshu, and K. Marsh. 1996. Deep breathing in children with severe malaria: indicator of metabolic acidosis and poor outcome. *Am J Trop Med Hyg* 55:521-524.

152. de Souza, J. B., and E. M. Riley. 2002. Cerebral malaria: the contribution of studies in animal models to our understanding of immunopathogenesis. *Microbes Infect* 4:291-300.

153. de Kossodo, S., and G. E. Grau. 1993. Profiles of cytokine production in relation with susceptibility to cerebral malaria. *J Immunol* 151:4811-4820.

154. Jennings, V. M., A. A. Lal, and R. L. Hunter. 1998. Evidence for multiple pathologic and protective mechanisms of murine cerebral malaria. *Infect Immun* 66:5972-5979.

155. Hearn, J., N. Rayment, D. N. Landon, D. R. Katz, and J. B. de Souza. 2000. Immunopathology of cerebral malaria: morphological evidence of parasite sequestration in murine brain microvasculature. *Infect Immun* 68:5364-5376.

156. Kaul, D. K., R. L. Nagel, J. F. Llena, and H. L. Shear. 1994. Cerebral malaria in mice: demonstration of cytoadherence of infected red blood cells and microrheologic correlates. *Am J Trop Med Hyg* 50:512-521.

157. Haque, A., H. Echchannaoui, R. Seguin, J. Schwartzman, L. H. Kasper, and S. Haque. 2001. Cerebral malaria in mice: interleukin-2 treatment induces accumulation of gammadelta T cells in the brain and alters resistant mice to susceptible-like phenotype. *Am J Pathol* 158:163-172.

158. Grau, G. E., L. F. Fajardo, P. F. Piguet, B. Allet, P. H. Lambert, and P. Vassalli. 1987. Tumor necrosis factor (cachectin) as an essential mediator in murine cerebral malaria. *Science* 237:1210-1212.

159. Rudin, W., H. P. Eugster, G. Bordmann, J. Bonato, M. Muller, M. Yamage, and B. Ryffel. 1997. Resistance to cerebral malaria in tumor necrosis factor-alpha/beta-deficient mice is associated with a reduction of intercellular adhesion molecule-1 up-regulation and T helper type 1 response. *Am J Pathol* 150:257-266.

160. Finley, R. W., L. J. Mackey, and P. H. Lambert. 1982. Virulent P. berghei malaria: prolonged survival and decreased cerebral pathology in cell-dependent nude mice. *J Immunol* 129:2213-2218.

161. Yanez, D. M., D. D. Manning, A. J. Cooley, W. P. Weidanz, and H. C. van der Heyde. 1996. Participation of lymphocyte subpopulations in the pathogenesis of experimental murine cerebral malaria. *J Immunol* 157:1620-1624.

162. Belnoue, E., M. Kayibanda, A. M. Vigario, J. C. Deschemin, N. van Rooijen, M. Viguier, G. Snounou, and L. Renia. 2002. On the pathogenic role of brain-sequestered alphabeta CD8+ T cells in experimental cerebral malaria. *J Immunol* 169:6369-6375.

163. Nitcheu, J., O. Bonduelle, C. Combadiere, M. Tefit, D. Seilhean, D. Mazier, and B. Combadiere. 2003. Perforin-dependent brain-infiltrating cytotoxic CD8+ T lymphocytes mediate experimental cerebral malaria pathogenesis. *J Immunol* 170:2221-2228.

164. Belnoue, E., M. Kayibanda, J. C. Deschemin, M. Viguier, M. Mack, W. A. Kuziel, and L. Renia. 2003. CCR5 deficiency decreases susceptibility to experimental cerebral malaria. *Blood* 101:4253-4259.

165. Chen, L., Z. Zhang, and F. Sendo. 2000. Neutrophils play a critical role in the pathogenesis of experimental cerebral malaria. *Clin Exp Immunol* 120:125-133.

166. Senaldi, G., C. Vesin, R. Chang, G. E. Grau, and P. F. Piguet. 1994. Role of polymorphonuclear neutrophil leukocytes and their integrin CD11a (LFA-1) in the pathogenesis of severe murine malaria. *Infect Immun* 62:1144-1149.

167. Chen, L., and F. Sendo. 2001. Cytokine and chemokine mRNA expression in neutrophils from CBA/NSlc mice infected with Plasmodium berghei ANKA that induces experimental cerebral malaria. *Parasitol Int* 50:139-143.

168. Medana, I. M., N. H. Hunt, and G. Chaudhri. 1997. Tumor necrosis factor-alpha expression in the brain during fatal murine cerebral malaria: evidence for production by microglia and astrocytes. *Am J Pathol* 150:1473-1486.

169. Hanum, P. S., M. Hayano, and S. Kojima. 2003. Cytokine and chemokine responses in a cerebral malaria-susceptible or -resistant strain of mice to Plasmodium berghei ANKA infection: early chemokine expression in the brain. *Int Immunol* 15:633-640.

170. Phillips, R. E., and G. Pasvol. 1992. Anaemia of Plasmodium falciparum malaria. *Baillieres Clin Haematol* 5:315-330.

171. Abdalla, S., D. J. Weatherall, S. N. Wickramasinghe, and M. Hughes. 1980. The anaemia of P. falciparum malaria. *Br J Haematol* 46:171-183.

172. Dormer, P., M. Dietrich, P. Kern, and R. D. Horstmann. 1983. Ineffective erythropoiesis in acute human P. falciparum malaria. *Blut* 46:279-288.

173. Burchard, G. D., P. Radloff, J. Philipps, M. Nkeyi, J. Knobloch, and P. G. Kremsner. 1995. Increased erythropoietin production in children with severe malarial anemia. *Am J Trop Med Hyg* 53:547-551.

174. Maggio-Price, L., D. Brookoff, and L. Weiss. 1985. Changes in hematopoietic stem cells in bone marrow of mice with Plasmodium berghei malaria. *Blood* 66:1080-1085.

175. Silverman, P. H., J. C. Schooley, and L. J. Mahlmann. 1987. Murine malaria decreases hemopoietic stem cells. *Blood* 69:408-413.

176. Clark, I. A. 1987. Monokines and lymphokines in malarial pathology. *Ann Trop Med Parasitol* 81:577-585.

177. Miller, K. L., J. C. Schooley, K. L. Smith, B. Kullgren, L. J. Mahlmann, and P. H. Silverman. 1989. Inhibition of erythropoiesis by a soluble factor in murine malaria. *Exp Hematol* 17:379-385.

178. Sam, H., and M. M. Stevenson. 1999. In vivo IL-12 production and IL-12 receptors beta1 and beta2 mRNA expression in the spleen are differentially up-regulated in resistant B6 and susceptible A/J mice during early blood-stage Plasmodium chabaudi AS malaria. *J Immunol* 162:1582-1589.

179. Yap, G. S., and M. M. Stevenson. 1994. Inhibition of in vitro erythropoiesis by soluble mediators in Plasmodium chabaudi AS malaria: lack of a major role for interleukin 1, tumor necrosis factor alpha, and gamma interferon. *Infect Immun* 62:357-362.

180. Martiney, J. A., B. Sherry, C. N. Metz, M. Espinoza, A. S. Ferrer, T. Calandra, H. E. Broxmeyer, and R. Bucala. 2000. Macrophage migration inhibitory factor release by macrophages after ingestion of Plasmodium chabaudi-infected erythrocytes: possible role in the pathogenesis of malarial anemia. *Infect Immun* 68:2259-2267.

181. Granja, A. C., F. Machungo, A. Gomes, S. Bergstrom, and B. Brabin. 1998. Malaria-related maternal mortality in urban Mozambique. *Ann Trop Med Parasitol* 92:257-263.

182. McGregor, I. A. 1984. Epidemiology, malaria and pregnancy. *Am J Trop Med Hyg* 33:517-525.

183. Leopardi, O., W. Naughten, L. Salvia, M. Colecchia, A. Matteelli, A. Zucchi, A. Shein, J. A. Muchi, G. Carosi, and M. Ghione. 1996. Malaric placentas. A quantitative study and clinico-pathological correlations. *Pathol Res Pract* 192:892-898; discussion 899-900.

184. Beeson, J. G., S. J. Rogerson, B. M. Cooke, J. C. Reeder, W. Chai, A. M. Lawson, M. E. Molyneux, and G. V. Brown. 2000. Adhesion of Plasmodium falciparum-infected erythrocytes to hyaluronic acid in placental malaria. *Nat Med* 6:86-90.

185. Ordi, J., M. R. Ismail, P. J. Ventura, E. Kahigwa, R. Hirt, A. Cardesa, P. L. Alonso, and C. Menendez. 1998. Massive chronic intervillositis of the placenta associated with malaria infection. *Am J Surg Pathol* 22:1006-1011.

186. Abrams, E. T., H. Brown, S. W. Chensue, G. D. Turner, E. Tadesse, V. M. Lema, M. E. Molyneux, R. Rochford, S. R. Meshnick, and S. J. Rogerson. 2003. Host response to malaria during pregnancy: placental monocyte recruitment is associated with elevated beta chemokine expression. *J Immunol* 170:2759-2764.
187. Ordi, J., C. Menendez, M. R. Ismail, P. J. Ventura, A. Palacin, E. Kahigwa, B. Ferrer, A. Cardesa, and P. L. Alonso. 2001. Placental malaria is associated with cell-mediated inflammatory responses with selective absence of natural killer cells. *J Infect Dis* 183:1100-1107.
188. Moore, J. M., B. L. Nahlen, A. Misore, A. A. Lal, and V. Udhayakumar. 1999. Immunity to placental malaria. I. Elevated production of interferon-gamma by placental blood mononuclear cells is associated with protection in an area with high transmission of malaria. *J Infect Dis* 179:1218-1225.
189. Fried, M., R. O. Muga, A. O. Misore, and P. E. Duffy. 1998. Malaria elicits type 1 cytokines in the human placenta: IFN-gamma and TNF-alpha associated with pregnancy outcomes. *J Immunol* 160:2523-2530.
190. Moormann, A. M., A. D. Sullivan, R. A. Rochford, S. W. Chensue, P. J. Bock, T. Nyirenda, and S. R. Meshnick. 1999. Malaria and pregnancy: placental cytokine expression and its relationship to intrauterine growth retardation. *J Infect Dis* 180:1987 1993.
191. Nosten, F., R. McGready, J. A. Simpson, K. L. Thwai, S. Balkan, T. Cho, L. Hkirijaroen, S. Looareesuwan, and N. J. White. 1999. Effects of Plasmodium vivax malaria in pregnancy. *Lancet* 354:546-549.
192. Tkachuk, A. N., A. M. Moormann, J. A. Poore, R. A. Rochford, S. W. Chensue, V. Mwapasa, and S. R. Meshnick. 2001. Malaria enhances expression of CC chemokine receptor 5 on placental macrophages. *J Infect Dis* 183:967-972.
193. Dragic, T., V. Litwin, G. P. Allaway, S. R. Martin, Y. Huang, K. A. Nagashima, C. Cayanan, P. J. Maddon, R. A. Koup, J. P. Moore, and W. A. Paxton. 1996. HIV-1 entry into CD4+ cells is mediated by the chemokine receptor CC-CKR-5. *Nature* 381:667-673.
194. Pichyangkul, S., P. Saengkrai, K. Yongvanitchit, D. G. Heppner, D. E. Kyle, and H. K. Webster. 1997. Regulation of leukocyte adhesion molecules CD11b/CD18 and leukocyte adhesion molecule-1 on phagocytic cells activated by malaria pigment. *Am J Trop Med Hyg* 57:383-388.
195. Schofield, L., S. Novakovic, P. Gerold, R. T. Schwarz, M. J. McConville, and S. D. Tachado. 1996. Glycosylphosphatidylinositol toxin of Plasmodium up-regulates intercellular adhesion molecule-1, vascular cell adhesion molecule-1, and E-selectin expression in vascular endothelial cells and increases leukocyte and parasite cytoadherence via tyrosine kinase-dependent signal transduction. *J Immunol* 156:1886-1896.
196. Prada, J., J. Malinowski, S. Muller, U. Bienzle, and P. G. Kremsner. 1995. Hemozoin differentially modulates the production of interleukin 6 and tumor necrosis factor in murine malaria. *Eur Cytokine Netw* 6:109-112.
197. Prada, J., J. Malinowski, S. Muller, U. Bienzle, and P. G. Kremsner. 1996. Effects of Plasmodium vinckei hemozoin on the production of oxygen radicals and nitrogen oxides in murine macrophages. *Am J Trop Med Hyg* 54:620-624.
198. Mordmuller, B., F. Turrini, H. Long, P. G. Kremsner, and P. Arese. 1998. Neutrophils and monocytes from subjects with the Mediterranean G6PD variant: effect of Plasmodium falciparum hemozoin on G6PD activity, oxidative burst and cytokine production. *Eur Cytokine Netw* 9:239-245.
199. Schwarzer, E., F. Turrini, G. Giribaldi, M. Cappadoro, and P. Arese. 1993. Phagocytosis of P. falciparum malarial pigment hemozoin by human monocytes inactivates monocyte protein kinase C. *Biochim Biophys Acta* 1181:51-54.
200. Schwarzer, E., and P. Arese. 1996. Phagocytosis of malarial pigment hemozoin inhibits NADPH-oxidase activity in human monocyte-derived macrophages. *Biochim Biophys Acta* 1316:169-175.
201. Schwarzer, E., H. Kuhn, E. Valente, and P. Arese. 2003. Malaria-parasitized erythrocytes and hemozoin nonenzymatically generate large amounts of hydroxy fatty acids that inhibit monocyte functions. *Blood* 101:722-728.
202. Perkins, D. J., J. M. Moore, J. Otieno, Y. P. Shi, B. L. Nahlen, V. Udhayakumar, and A. A. Lal. 2003. In vivo acquisition of hemozoin by placental blood mononuclear cells suppresses PGE2, TNF-alpha, and IL-10. *Biochem Biophys Res Commun* 311:839-846.
203. Taramelli, D., N. Basilico, A. M. De Palma, M. Saresella, P. Ferrante, L. Mussoni, and P. Olliaro. 1998. The effect of synthetic malaria pigment (beta-haematin) on adhesion molecule expression and interleukin-6 production by human endothelial cells. *Trans R Soc Trop Med Hyg* 92:57-62.
204. Taramelli, D., S. Recalcati, N. Basilico, P. Olliaro, and G. Cairo. 2000. Macrophage preconditioning with synthetic malaria pigment reduces cytokine production via heme iron-dependent oxidative stress. *Lab Invest* 80:1781-1788.
205. Biswas, S., M. G. Karmarkar, and Y. D. Sharma. 2001. Antibodies detected against Plasmodium falciparum haemozoin with inhibitory properties to cytokine production. *FEMS Microbiol Lett* 194:175-179.
206. Gowda, D. C. 2002. Structure and activity of glycosylphosphatidylinositol anchors of Plasmodium falciparum. *Microbes Infect* 4:983-990.
207. Ferguson, M. A. 1994. What can GPI do for you? *Parasitol Today* 10:48-52.
208. Braun Breton, C., T. L. Rosenberry, and L. H. Pereira da Silva. 1990. Glycolipid anchorage of Plasmodium falciparum surface antigens. *Res Immunol* 141:743-755.

209. Naik, R. S., E. A. Davidson, and D. C. Gowda. 2000. Developmental stage-specific biosynthesis of glycosylphosphatidylinositol anchors in intraerythrocytic Plasmodium falciparum and its inhibition in a novel manner by mannosamine. *J Biol Chem* 275:24506-24511.

210. Gowda, D. C., P. Gupta, and E. A. Davidson. 1997. Glycosylphosphatidylinositol anchors represent the major carbohydrate modification in proteins of intraerythrocytic stage Plasmodium falciparum. *J Biol Chem* 272:6428-6439.

211. Gerold, P., L. Schofield, M. J. Blackman, A. A. Holder, and R. T. Schwarz. 1996. Structural analysis of the glycosyl-phosphatidylinositol membrane anchor of the merozoite surface proteins-1 and -2 of Plasmodium falciparum. *Mol Biochem Parasitol* 75:131-143.

212. Berhe, S., L. Schofield, R. T. Schwarz, and P. Gerold. 1999. Conservation of structure among glycosylphosphatidylinositol toxins from different geographic isolates of Plasmodium falciparum. *Mol Biochem Parasitol* 103:273-278.

213. Gerold, P., L. Vivas, S. A. Ogun, N. Azzouz, K. N. Brown, A. A. Holder, and R. T. Schwarz. 1997. Glycosylphosphatidylinositols of Plasmodium chabaudi chabaudi: a basis for the study of malarial glycolipid toxins in a rodent model. *Biochem J* 328 (Pt 3):905-911.

214. Kimmel, J., S. A. Ogun, C. S. de Macedo, P. Gerold, L. Vivas, A. A. Holder, R. T. Schwarz, and N. Azzouz. 2003. Glycosylphosphatidyl-inositols in murine malaria: Plasmodium yoelii yoelii. *Biochimie* 85:473-481.

215. Kooyman, D. L., G. W. Byrne, S. McClellan, D. Nielsen, M. Tone, H. Waldmann, T. M. Coffman, K. R. McCurry, J. L. Platt, and J. S. Logan. 1995. In vivo transfer of GPI-linked complement restriction factors from erythrocytes to the endothelium. *Science* 269:89-92.

216. Rooney, I. A., J. E. Heuser, and J. P. Atkinson. 1996. GPI-anchored complement regulatory proteins in seminal plasma. An analysis of their physical condition and the mechanisms of their binding to exogenous cells. *J Clin Invest* 97:1675-1686.

217. Camargo, M. M., A. C. Andrade, I. C. Almeida, L. R. Travassos, and R. T. Gazzinelli. 1997. Glycoconjugates isolated from Trypanosoma cruzi but not from Leishmania species membranes trigger nitric oxide synthesis as well as microbicidal activity in IFN-gamma-primed macrophages. *J Immunol* 159:6131-6139.

218. Almeida, I. C., M. M. Camargo, D. O. Procopio, L. S. Silva, A. Mehlert, L. R. Travassos, R. T. Gazzinelli, and M. A. Ferguson. 2000. Highly purified glycosylphosphatidylinositols from Trypanosoma cruzi are potent proinflammatory agents. *EMBO J* 19:1476-1485.

219. Bate, C. A., J. Taverne, N. D. Karunaweera, K. N. Mendis, D. Kwiatkowski, and J. H. Playfair. 1992. Serological relationship of tumor necrosis factor-inducing exoantigens of Plasmodium falciparum and Plasmodium vivax. *Infect Immun* 60:1241-1243.

220. Bate, C. A., J. Taverne, and J. H. Playfair. 1989. Soluble malarial antigens are toxic and induce the production of tumour necrosis factor in vivo. *Immunology* 66:600-605.

221. Tachado, S. D., and L. Schofield. 1994. Glycosylphosphatidylinositol toxin of Trypanosoma brucei regulates IL-1 alpha and TNF-alpha expression in macrophages by protein tyrosine kinase mediated signal transduction. *Biochem Biophys Res Commun* 205:984-991.

222. Magez, S., B. Stijlemans, M. Radwanska, E. Pays, M. A. Ferguson, and P. De Baetselier. 1998. The glycosyl-inositol-phosphate and dimyristoylglycerol moieties of the glycosylphosphatidylinositol anchor of the trypanosome variant-specific surface glycoprotein are distinct macrophage-activating factors. *J Immunol* 160:1949-1956.

223. Proudfoot, L., C. A. O'Donnell, and F. Y. Liew. 1995. Glycoinositolphospholipids of Leishmania major inhibit nitric oxide synthesis and reduce leishmanicidal activity in murine macrophages. *Eur J Immunol* 25:745-750.

224. Chawla, M., and R. A. Vishwakarma. 2003. Alkylacylglycerolipid domain of GPI molecules of Leishmania is responsible for inhibition of PKC-mediated c-fos expression. *J Lipid Res* 44:594-600.

225. Tachado, S. D., P. Gerold, R. Schwarz, S. Novakovic, M. McConville, and L. Schofield. 1997. Signal transduction in macrophages by glycosylphosphatidylinositols of Plasmodium, Trypanosoma, and Leishmania: activation of protein tyrosine kinases and protein kinase C by inositolglycan and diacylglycerol moieties. *Proc Natl Acad Sci U S A* 94:4022-4027.

226. Vijaykumar, M., R. S. Naik, and D. C. Gowda. 2001. Plasmodium falciparum glycosylphosphatidylinositol-induced TNF-alpha secretion by macrophages is mediated without membrane insertion or endocytosis. *J Biol Chem* 276:6909-6912.

227. Naik, R. S., O. H. Branch, A. S. Woods, M. Vijaykumar, D. J. Perkins, B. L. Nahlen, A. A. Lal, R. J. Cotter, C. E. Costello, C. F. Ockenhouse, E. A. Davidson, and D. C. Gowda. 2000. Glycosylphosphatidylinositol anchors of Plasmodium falciparum: molecular characterization and naturally elicited antibody response that may provide immunity to malaria pathogenesis. *J Exp Med* 192:1563-1576.

228. Schofield, L. 2002. Antidisease vaccines. *Chem Immunol* 80:322-342.

229. Schofield, L., M. C. Hewitt, K. Evans, M. A. Siomos, and P. H. Seeberger. 2002. Synthetic GPI as a candidate anti-toxic vaccine in a model of malaria. *Nature* 418:785-789.

230. Sinton, J. A., and B. N. Ghosh. 1934. Studies of malarial pigment (hemozoin). Part I. Investigation of the action of solvents on hemozoin and the spectroscopical appearances observed in the solutions. *Rec. Malaria Surv. India.*

231. Wernsdorfer, W. H., and I. E. McGregor. 1988. *Malaria: Principles and Practice of Malarology*. Churchill-Livingstone, Edinburgh.

232. Meckel, H. 1847. Uber schwartzes pigment in der milz und dem blute einer geisteskranken. *Zeitsch. Psychiat.* 4:198.

233. Virchow, R. 1849. Zur pathologischen physiologie des blutes. *Virchow's Arch. Pathol. Anat. Physiol.* 2:587.

234. Hischl, E. 1850. Uber pigmentbildung nach febris intermitten. . *Zeitschrift der k.k. Gesellschaft der aertzte in Wein*:338.

235. Laveran, A. 1880. Un nouveau parasite trouve dans le sang des malades atteints de fievre palustre. Origine parasitaire des accidents de l'impaludisme. *Bull. Acad. Med.* 9:1235.

236. Councilman, W. T., and A. C. Abbot. 1885. A contribution to the pathology of malaria fever. *Am. J. Med. Sci.* 89:416.

237. Ross, R. 1897. On some peculiar pigmented cells found in two mosquitos fed on malarial blood. *Br. Med. J.* 1:1786.

238. Brown, W. H. 1911. Malarial pigment (so-called melanin): its nature and mode of production. *J. Exp. Med.* 13:290.

239. Brown, W. H. 1913. Malarial pigment (hematin) as an active factor in the production of the blood picture of malaria. *J. Exp. Med.* 18:96.

240. Ghosh, B. N., and M. C. Nath. 1934. The chemical composition of malaria pigment (hemozoin). *Rec. Malaria Surv. India* 4:321.

241. Ghosh, B. N., and J. A. Sinton. 1934. Studies of malaria pigment (hemozoin). The reactions of HZ to tests for iron. *Rec. Malaria Surv. India* 4:43.

242. Deegan, T., and B. G. Maegraith. 1956. Studies on the nature of malarial pigment (haemozion). I. The pigment of the simian species, Plasmodium knowlesi and P. Cynomolgi. *Ann Trop Med Parasitol* 50:194-211.

243. Sherman, I. W., I. P. Ting, and J. A. Ruble. 1968. Characterization of the malaria pigment (hemozoin) from the avian malaria parasite Plasmodium lophurae. *J Protozool* 15:158-164.

244. Yamada, K. A., and I. W. Sherman. 1979. Plasmodium lophurae: composition and properties of hemozoin, the malarial pigment. *Exp Parasitol* 48:61-74.

245. Fitch, C. D., and P. Kanjananggulpan. 1987. The state of ferriprotoporphyrin IX in malaria pigment. *J Biol Chem* 262:15552-15555.

246. Slater, A. F., W. J. Swiggard, B. R. Orton, W. D. Flitter, D. E. Goldberg, A. Cerami, and G. B. Henderson. 1991. An iron-carboxylate bond links the heme units of malaria pigment. *Proc Natl Acad Sci U S A* 88:325-329.

247. Slater, A. F. 1992. Malaria pigment. *Exp Parasitol* 74:362-365.

248. Bohle, D. S., R. E. Dinnebier, S. K. Madsen, and P. W. Stephens. 1997. Characterization of the products of the heme detoxification pathway in malarial late trophozoites by X-ray diffraction. *J Biol Chem* 272:713-716.

249. Pagola, S., P. W. Stephens, D. S. Bohle, A. D. Kosar, and S. K. Madsen. 2000. The structure of malaria pigment beta-haematin. *Nature* 404:307-310.

250. Bohle, D. S., A. D. Kosar, and P. W. Stephens. 2002. Phase homogeneity and crystal morphology of the malaria pigment beta-hematin. *Acta Crystallogr D Biol Crystallogr* 58:1752-1756.

251. Egan, T. J., W. W. Mavuso, and K. K. Ncokazi. 2001. The mechanism of beta-hematin formation in acetate solution. Parallels between hemozoin formation and biomineralization processes. *Biochemistry* 40:204-213.

252. Sherman, I. W. 1977. Amino acid metabolism and protein synthesis in malarial parasites. *Bull World Health Organ* 55:265-276.

253. Polet, H., and M. E. Conrad. 1968. Malaria: extracellular amino acid requirements for in vitro growth of erythrocytic forms of Plasmodium knowlesi. *Proc Soc Exp Biol Med* 127:251-253.

254. Rudzinska, M. A., W. Trager, and R. S. Bray. 1965. Pinocytotic uptake and the digestion of hemoglobin in malaria parasites. *J Protozool* 12:563-576.

255. Yayon, A., Z. I. Cabantchik, and H. Ginsburg. 1984. Identification of the acidic compartment of Plasmodium falciparum-infected human erythrocytes as the target of the antimalarial drug chloroquine. *EMBO J* 3:2695-2700.

256. Sherman, I. W. 1979. Biochemistry of Plasmodium (malarial parasites). *Microbiol Rev* 43:453-495.

257. Orjih, A. U., and C. D. Fitch. 1993. Hemozoin production by Plasmodium falciparum: variation with strain and exposure to chloroquine. *Biochim Biophys Acta* 1157:270-274.

258. Zarchin, S., M. Krugliak, and H. Ginsburg. 1986. Digestion of the host erythrocyte by malaria parasites is the primary target for quinoline-containing antimalarials. *Biochem Pharmacol* 35:2435-2442.

259. Hempelmann, E., and R. J. Wilson. 1980. Endopeptidases from Plasmodium knowlesi. *Parasitology* 80:323-330.

260. Vander Jagt, D. L., L. A. Hunsaker, and N. M. Campos. 1986. Characterization of a hemoglobin-degrading, low molecular weight protease from Plasmodium falciparum. *Mol Biochem Parasitol* 18:389-400.

261. Rosenthal, P. J., J. H. McKerrow, M. Aikawa, H. Nagasawa, and J. H. Leech. 1988. A malarial cysteine proteinase is necessary for hemoglobin degradation by Plasmodium falciparum. *J Clin Invest* 82:1560-1566.

262. Sherman, I. W., and L. Tanigoshi. 1983. Purification of Plasmodium lophurae cathepsin D and its effects on erythrocyte membrane proteins. *Mol Biochem Parasitol* 8:207-226.

263. Goldberg, D. E., A. F. Slater, A. Cerami, and G. B. Henderson. 1990. Hemoglobin degradation in the malaria parasite Plasmodium falciparum: an ordered process in a unique organelle. *Proc Natl Acad Sci U S A* 87:2931-2935.

264. Gluzman, I. Y., S. E. Francis, A. Oksman, C. E. Smith, K. L. Duffin, and D. E. Goldberg. 1994. Order and specificity of the Plasmodium falciparum hemoglobin degradation pathway. *J Clin Invest* 93:1602-1608.

265. Francis, S. E., D. J. Sullivan, Jr., and D. E. Goldberg. 1997. Hemoglobin metabolism in the malaria parasite Plasmodium falciparum. *Annu Rev Microbiol* 51:97-123.

266. Salas, F., J. Fichmann, G. K. Lee, M. D. Scott, and P. J. Rosenthal. 1995. Functional expression of falcipain, a Plasmodium falciparum cysteine proteinase, supports its role as a malarial hemoglobinase. *Infect Immun* 63:2120-2125.

267. Goldberg, D. E. 1993. Hemoglobin degradation in Plasmodium-infected red blood cells. *Semin Cell Biol* 4:355-361.

268. Slomianny, C. 1990. Three-dimensional reconstruction of the feeding process of the malaria parasite. *Blood Cells* 16:369-378.

269. Yayon, A., R. Timberg, S. Friedman, and H. Ginsburg. 1984. Effects of chloroquine on the feeding mechanism of the intraerythrocytic human malarial parasite Plasmodium falciparum. *J Protozool* 31:367-372.

270. Olliaro, P., F. Castelli, F. Milano, G. Filice, and G. Carosi. 1989. Ultrastructure of Plasmodium falciparum "in vitro". I. Base-line for drug effects evaluation. *Microbiologica* 12:7-14.

271. Hempelmann, E., C. Motta, R. Hughes, S. A. Ward, and P. G. Bray. 2003. Plasmodium falciparum: sacrificing membrane to grow crystals? *Trends Parasitol* 19:23-26.

272. Ginsburg, H. 2003. The mysteries of hemoglobin degradation in Plasmodium falciparum-infected erythrocytes. *Trends Parasitol* 19:198-199; author reply 199-200.

273. Ziegler, J., R. Linck, and D. W. Wright. 2001. Heme Aggregation inhibitors: antimalarial drugs targeting an essential biomineralization process. *Curr Med Chem* 8:171-189.

274. Yasuhara, T., M. Mori, K. Wakamatsu, and K. Kubo. 1991. Isolation and identification of hemin as an endogenous Na+/K(+)-ATPase inhibitor from porcine blood cells. *Biochem Biophys Res Commun* 178:95-103.

275. Tappel, A. L. 1953. The mechanism of the oxidation of unsaturated fatty acids catalyzed by hematin compounds. *Arch Biochem Biophys* 44:378-395.

276. Atamna, H., and H. Ginsburg. 1993. Origin of reactive oxygen species in erythrocytes infected with Plasmodium falciparum. *Mol Biochem Parasitol* 61:231-241.

277. Goldberg, D. E. 1992. Plasmodial hemoglobin degradation: an ordered pathway in a specialized organelle. *Infect Agents Dis* 1:207-211.

278. Hempelmann, E., and T. J. Egan. 2002. Pigment biocrystallization in Plasmodium falciparum. *Trends Parasitol* 18:11.

279. Egan, T. J. 2002. Physico-chemical aspects of hemozoin (malaria pigment) structure and formation. *J Inorg Biochem* 91:19-26.

280. Krogstad, D. J., P. H. Schlesinger, and I. Y. Gluzman. 1985. Antimalarials increase vesicle pH in Plasmodium falciparum. *J Cell Biol* 101:2302-2309.

281. Chou, A. C., and C. D. Fitch. 1992. Heme polymerase: modulation by chloroquine treatment of a rodent malaria. *Life Sci* 51:2073-2078.

282. Egan, T. J., D. C. Ross, and P. A. Adams. 1994. Quinoline anti-malarial drugs inhibit spontaneous formation of beta-haematin (malaria pigment). *FEBS Lett* 352:54-57.

283. Dorn, A., R. Stoffel, H. Matile, A. Bubendorf, and R. G. Ridley. 1995. Malarial haemozoin/beta-haematin supports haem polymerization in the absence of protein. *Nature* 374:269-271.

284. Adams, P. A., T. J. Egan, D. C. Ross, J. Silver, and P. J. Marsh. 1996. The chemical mechanism of beta-haematin formation studied by Mossbauer spectroscopy. *Biochem J* 318 (Pt 1):25-27.

285. Pandey, A. V., and B. L. Tekwani. 1996. Formation of haemozoin/beta-haematin under physiological conditions is not spontaneous. *FEBS Lett* 393:189-193.

286. Fitch, C. D., and A. C. Chou. 1996. Heat-labile and heat-stimulable heme polymerase activities in Plasmodium berghei. *Mol Biochem Parasitol* 82:261-264.

287. Pandey, A. V., N. Singh, B. L. Tekwani, S. K. Puri, and V. S. Chauhan. 1999. Assay of beta-hematin formation by malaria parasite. *J Pharm Biomed Anal* 20:203-207.

288. Sullivan, D. J., Jr., I. Y. Gluzman, and D. E. Goldberg. 1996. Plasmodium hemozoin formation mediated by histidine-rich proteins. *Science* 271:219-222.

289. Bendrat, K., B. J. Berger, and A. Cerami. 1995. Haem polymerization in malaria. *Nature* 378:138-139.

290. Lynn, A., S. Chandra, P. Malhotra, and V. S. Chauhan. 1999. Heme binding and polymerization by Plasmodium falciparum histidine rich protein II: influence of pH on activity and conformation. *FEBS Lett* 459:267-271.

291. Papalexis, V., M. A. Siomos, N. Campanale, X. Guo, G. Kocak, M. Foley, and L. Tilley. 2001. Histidine-rich protein 2 of the malaria parasite, Plasmodium falciparum, is involved in detoxification of the by-products of haemoglobin degradation. *Mol Biochem Parasitol* 115:77-86.
292. Noland, G. S., N. Briones, and D. J. Sullivan, Jr. 2003. The shape and size of hemozoin crystals distinguishes diverse Plasmodium species. *Mol Biochem Parasitol* 130:91-99.
293. Ziegler, J. R., T. Chang, and D. W. Wright. 1999. Multiple-Antigenic Peptides of Histidine-Rich Protein II of Plasmodium falciparum: Dendrimeric Biomineralization Templates. *J. Am. Chem. Soc* 121:2395.
294. Pandey, A. V., R. Joshi, B. L. Tekwani, R. L. Singh, and V. S. Chauhan. 1997. Synthetic peptides corresponding to a repetitive sequence of malarial histidine rich protein bind haem and inhibit haemozoin formation in vitro. *Mol Biochem Parasitol* 90:281-287.
295. Dorn, A., S. R. Vippagunta, H. Matile, A. Bubendorf, J. L. Vennerstrom, and R. G. Ridley. 1998. A comparison and analysis of several ways to promote haematin (haem) polymerisation and an assessment of its initiation in vitro. *Biochem Pharmacol* 55:737-747.
296. Fitch, C. D., G. Z. Cai, Y. F. Chen, and J. D. Shoemaker. 1999. Involvement of lipids in ferriprotoporphyrin IX polymerization in malaria. *Biochim Biophys Acta* 1454:31-37.
297. Pandey, A. V., V. K. Babbarwal, J. N. Okoyeh, R. M. Joshi, S. K. Puri, R. L. Singh, and V. S. Chauhan. 2003. Hemozoin formation in malaria: a two-step process involving histidine-rich proteins and lipids. *Biochem Biophys Res Commun* 308:736-743.
298. Bazylinski, D. A., A. J. Garratt-Reed, and R. B. Frankel. 1994. Electron microscopic studies of magnetosomes in magnetotactic bacteria. *Microsc Res Tech* 27:389-401.
299. Wood, P. A., and J. W. Eaton. 1993. Hemoglobin catabolism and host-parasite heme balance in chloroquine-sensitive and chloroquine-resistant Plasmodium berghei infections. *Am J Trop Med Hyg* 48:465-472.
300. Oliveira, M. F., J. R. Silva, M. Dansa-Petretski, W. de Souza, U. Lins, C. M. Braga, H. Masuda, and P. L. Oliveira. 1999. Haem detoxification by an insect. *Nature* 400:517-518.
301. Lehane, M. J. 1991. *Biology of Blood-Sucking Insects*. Harper Collins Academic. London.
302. Dansa-Petretski, M., J. M. Ribeiro, G. C. Atella, H. Masuda, and P. L. Oliveira. 1995. Antioxidant role of Rhodnius prolixus heme-binding protein. Protection against heme-induced lipid peroxidation. *J Biol Chem* 270:10893-10896.
303. Oliveira, M. F., J. R. Silva, M. Dansa-Petretski, W. de Souza, C. M. Braga, H. Masuda, and P. L. Oliveira. 2000. Haemozoin formation in the midgut of the blood-sucking insect Rhodnius prolixus. *FEBS Lett* 477:95-98.
304. Brindley, P. J., B. H. Kalinna, J. P. Dalton, S. R. Day, J. Y. Wong, M. L. Smythe, and D. P. McManus. 1997. Proteolytic degradation of host hemoglobin by schistosomes. *Mol Biochem Parasitol* 89:1-9.
305. Homewood, C. A., J. M. Jewsbury, and M. L. Chance. 1972. The pigment formed during haemoglobin digestion by malarial and schistosomal parasites. *Comp Biochem Physiol B* 43:517-523.
306. Moore, G. A., C. A. Homewood, and H. M. Gilles. 1975. A comparison of pigment from Schistosoma mansoni and Plasmodium berghei. *Ann Trop Med Parasitol* 69:373-374.
307. Oliveira, M. F., J. C. d'Avila, C. R. Torres, P. L. Oliveira, A. J. Tempone, F. D. Rumjanek, C. M. Braga, J. R. Silva, M. Dansa-Petretski, M. A. Oliveira, W. de Souza, and S. T. Ferreira. 2000. Haemozoin in Schistosoma mansoni. *Mol Biochem Parasitol* 111:217-221.
308. Bradbury, P. C., and W. Trager. 1968. The fine structure of the mature gametes of haemoproteus columbae kruse. *J Protozool* 15:89-102.
309. Chen, M. M., L. Shi, and D. J. Sullivan, Jr. 2001. Haemoproteus and Schistosoma synthesize heme polymers similar to Plasmodium hemozoin and beta-hematin. *Mol Biochem Parasitol* 113:1-8.
310. Sullivan, D. J. 2002. Theories on malarial pigment formation and quinoline action. *Int J Parasitol* 32:1645-1653.
311. Langreth, S. G. 1976. Feeding mechanisms in extracellular Babesia microti and Plasmodium lophurae. *J Protozool* 23:215-223.
312. Halton, D. W. 1997. Nutritional adaptations to parasitism within the platyhelminthes. *Int J Parasitol* 27:693-704.
313. Bohle, D. S., B. J. Conklin, D. Cox, S. K. Madsen, S. Paulson, P. W. Stephens, and G. T. Yee. 1994. Structural and Spectroscopic Studies of beta-Hematin (the Heme Coordination Polymer in Malaria Pigment). In *Inorganic and Organometallic Polymers II. ACS Symp Series*. 497-515.
314. Blauer, G., and M. Akkawi. 1997. Investigations of B- and beta-hematin. *J Inorg Biochem* 66:145-152.
315. Egan, T. J., E. Hempelmann, and W. W. Mavuso. 1999. Characterisation of synthetic beta-haematin and effects of the antimalarial drugs quinidine, halofantrine, desbutylhalofantrine and mefloquine on its formation. *J Inorg Biochem* 73:101-107.
316. Blauer, G., and M. Akkawi. 2000. On the preparation of beta-haematin. *Biochem J* 346 Pt 2:249-250.
317. Bohle, D. S., and J. B. Helms. 1993. Synthesis of beta-hematin by dehydrohalogenation of hemin. *Biochem Biophys Res Commun* 193:504-508.
318. Rosenthal, P. J. 1995. Plasmodium falciparum: effects of proteinase inhibitors on globin hydrolysis by cultured malaria parasites. *Exp Parasitol* 80:272-281.

319. Bailly, E., R. Jambou, J. Savel, and G. Jaureguiberry. 1992. Plasmodium falciparum: differential sensitivity in vitro to E-64 (cysteine protease inhibitor) and Pepstatin A (aspartyl protease inhibitor). *J Protozool* 39:593-599.

320. Bray, P. G., M. Mungthin, R. G. Ridley, and S. A. Ward. 1998. Access to hematin: the basis of chloroquine resistance. *Mol Pharmacol* 54:170-179.

321. Francis, S. E., I. Y. Gluzman, A. Oksman, A. Knickerbocker, R. Mueller, M. L. Bryant, D. R. Sherman, D. G. Russell, and D. E. Goldberg. 1994. Molecular characterization and inhibition of a Plasmodium falciparum aspartic hemoglobinase. *EMBO J* 13:306-317.

322. Haque, T. S., A. G. Skillman, C. E. Lee, H. Habashita, I. Y. Gluzman, T. J. Ewing, D. E. Goldberg, I. D. Kuntz, and J. A. Ellman. 1999. Potent, low-molecular-weight non-peptide inhibitors of malarial aspartyl protease plasmepsin II. *J Med Chem* 42:1428-1440.

323. Li, R., G. L. Kenyon, F. E. Cohen, X. Chen, B. Gong, J. N. Dominguez, E. Davidson, G. Kurzban, R. E. Miller, E. O. Nuzum, and et al. 1995. In vitro antimalarial activity of chalcones and their derivatives. *J Med Chem* 38:5031-5037.

324. Rosenthal, P. J., J. E. Olson, G. K. Lee, J. T. Palmer, J. L. Klaus, and D. Rasnick. 1996. Antimalarial effects of vinyl sulfone cysteine proteinase inhibitors. *Antimicrob Agents Chemother* 40:1600-1603.

325. Olson, J. E., G. K. Lee, A. Semenov, and P. J. Rosenthal. 1999. Antimalarial effects in mice of orally administered peptidyl cysteine protease inhibitors. *Bioorg Med Chem* 7:633-638.

326. O'Neill, P. M., P. G. Bray, S. R. Hawley, S. A. Ward, and B. K. Park. 1998. 4-Aminoquinolines--past, present, and future: a chemical perspective. *Pharmacol Ther* 77:29-58.

327. W.H.O. 2001. Report No. W.H.O./CDS/ RBM/2001.33. World Health Organization. Geneva.

328. Ginsburg, H., S. A. Ward, and P. G. Bray. 1999. An integrated model of chloroquine action. *Parasitol Today* 15:357-360.

329. Krishna, S., and N. J. White. 1996. Pharmacokinetics of quinine, chloroquine and amodiaquine. Clinical implications. *Clin Pharmacokinet* 30:263-299.

330. Meshnick, S. R., and J. J. Marr. 1992. Intracellular parasites. New developments in chemotherapy. *Subcell Biochem* 18:401-441.

331. Slater, A. F. 1993. Chloroquine: mechanism of drug action and resistance in Plasmodium falciparum. *Pharmacol Ther* 57:203-235.

332. Sullivan, D. J., Jr., I. Y. Gluzman, D. G. Russell, and D. E. Goldberg. 1996. On the molecular mechanism of chloroquine's antimalarial action. *Proc Natl Acad Sci U S A* 93:11865-11870.

333. Moreau, S., B. Perly, C. Chachaty, and C. Deleuze. 1985. A nuclear magnetic resonance study of the interactions of antimalarial drugs with porphyrins. *Biochim Biophys Acta* 840:107-116.

334. Sullivan, D. J., Jr., H. Matile, R. G. Ridley, and D. E. Goldberg. 1998. A common mechanism for blockade of heme polymerization by antimalarial quinolines. *J Biol Chem* 273:31103-31107.

335. Sugioka, Y., M. Suzuki, K. Sugioka, and M. Nakano. 1987. A ferriprotoporphyrin IX-chloroquine complex promotes membrane phospholipid peroxidation. A possible mechanism for antimalarial action. *FEBS Lett* 223:251-254.

336. Chong, C. R., and D. J. Sullivan, Jr. 2003. Inhibition of heme crystal growth by antimalarials and other compounds: implications for drug discovery. *Biochem Pharmacol* 66:2201-2212.

337. Ginsburg, H., O. Famin, J. Zhang, and M. Krugliak. 1998. Inhibition of glutathione-dependent degradation of heme by chloroquine and amodiaquine as a possible basis for their antimalarial mode of action. *Biochem Pharmacol* 56:1305-1313.

338. Platel, D. F., F. Mangou, and J. Tribouley-Duret. 1999. Role of glutathione in the detoxification of ferriprotoporphyrin IX in chloroquine resistant Plasmodium berghei. *Mol Biochem Parasitol* 98:215-223.

339. Deharo, E., D. Barkan, M. Krugliak, J. Golenser, and H. Ginsburg. 2003. Potentiation of the antimalarial action of chloroquine in rodent malaria by drugs known to reduce cellular glutathione levels. *Biochem Pharmacol* 66:809-817.

340. Kwakye-Berko, F., and S. R. Meshnick. 1989. Binding of chloroquine to DNA. *Mol Biochem Parasitol* 35:51-55.

341. Surolia, N., and G. Padmanaban. 1991. Chloroquine inhibits heme-dependent protein synthesis in Plasmodium falciparum. *Proc Natl Acad Sci U S A* 88:4786-4790.

342. Gabay, T., M. Krugliak, G. Shalmiev, and H. Ginsburg. 1994. Inhibition by anti-malarial drugs of haemoglobin denaturation and iron release in acidified red blood cell lysates--a possible mechanism of their anti-malarial effect? *Parasitology* 108 (Pt 4):371-381.

343. Zidovetzki, R., I. W. Sherman, and L. O'Brien. 1993. Inhibition of Plasmodium falciparum phospholipase A2 by chloroquine, quinine, and arteether. *J Parasitol* 79:565-570.

344. Loria, P., S. Miller, M. Foley, and L. Tilley. 1999. Inhibition of the peroxidative degradation of haem as the basis of action of chloroquine and other quinoline antimalarials. *Biochem J* 339 (Pt 2):363-370.

345. Rosenthal, P. J., and S. R. Meshnick. 1996. Hemoglobin catabolism and iron utilization by malaria parasites. *Mol Biochem Parasitol* 83:131-139.

346. Olliaro, P. L., R. K. Haynes, B. Meunier, and Y. Yuthavong. 2001. Possible modes of action of the artemisinin-type compounds. *Trends Parasitol* 17:122-126.
347. Klayman, D. L. 1985. Qinghaosu (artemisinin): an antimalarial drug from China. *Science* 228:1049-1055.
348. Wu, Y. 2002. How might qinghaosu (artemisinin) and related compounds kill the intraerythrocytic malaria parasite? A chemist's view. *Acc Chem Res* 35:255-259.
349. Hong, Y. L., Y. Z. Yang, and S. R. Meshnick. 1994. The interaction of artemisinin with malarial hemozoin. *Mol Biochem Parasitol* 63:121-128.
350. Pandey, A. V., B. L. Tekwani, R. L. Singh, and V. S. Chauhan. 1999. Artemisinin, an endoperoxide antimalarial, disrupts the hemoglobin catabolism and heme detoxification systems in malarial parasite. *J Biol Chem* 274:19383-19388.
351. Meshnick, S. R., T. E. Taylor, and S. Kamchonwongpaisan. 1996. Artemisinin and the antimalarial endoperoxides: from herbal remedy to targeted chemotherapy. *Microbiol Rev* 60:301-315.
352. Akompong, T., J. VanWye, N. Ghori, and K. Haldar. 1999. Artemisinin and its derivatives are transported by a vacuolar-network of Plasmodium falciparum and their anti-malarial activities are additive with toxic sphingolipid analogues that block the network. *Mol Biochem Parasitol* 101:71-79.
353. Haynes, R. K., D. Monti, D. Taramelli, N. Basilico, S. Parapini, and P. Olliaro. 2003. Artemisinin antimalarials do not inhibit hemozoin formation. *Antimicrob Agents Chemother* 47:1175.
354. Asawamahasakda, W., I. Ittarat, C. C. Chang, P. McElroy, and S. R. Meshnick. 1994. Effects of antimalarials and protease inhibitors on plasmodial hemozoin production. *Mol Biochem Parasitol* 67:183-191.
355. Meshnick, S. R. 2003. Artemisinin and heme. *Antimicrob Agents Chemother* 47:2712; author reply 2712-2713.
356. Meshnick, S. R., Y. Z. Yang, V. Lima, F. Kuypers, S. Kamchonwongpaisan, and Y. Yuthavong. 1993. Iron-dependent free radical generation from the antimalarial agent artemisinin (qinghaosu). *Antimicrob Agents Chemother* 37:1108-1114.
357. Haynes, R. K. 2001. Artemisinin and derivatives: the future for malaria treatment? *Curr Opin Infect Dis* 14:719-726.
358. Basilico, N., D. Monti, P. Olliaro, and D. Taramelli. 1997. Non-iron porphyrins inhibit beta-haematin (malaria pigment) polymerisation. *FEBS Lett* 409:297-299.
359. Bhattacharjee, A. K., and J. M. Karle. 1996. Molecular electronic properties of a series of 4-quinolinecarbinolamines define antimalarial activity profile. *J Med Chem* 39:4622-4629.
360. Martiney, J. A., A. Cerami, and A. F. Slater. 1996. Inhibition of hemozoin formation in Plasmodium falciparum trophozoite extracts by heme analogs: possible implication in the resistance to malaria conferred by the beta-thalassemia trait. *Mol Med* 2:236-246.
361. Cannon, J. B., C. Martin, G. S. Drummond, and A. Kappas. 1993. Targeted delivery of a heme oxygenase inhibitor with a lyophilized liposomal tin mesoporphyrin formulation. *Pharm Res* 10:715-721.
362. Iyer, J. K., L. Shi, A. H. Shankar, and D. J. Sullivan, Jr. 2003. Zinc protoporphyrin IX binds heme crystals to inhibit the process of crystallization in Plasmodium falciparum. *Mol Med* 9:175-182.
363. Han, P., K. P. Fung, C. P. Teo, and L. P. Tam. 1990. Alpha-thalassaemia and beta-thalassaemia traits: biological difference based on red cell indices and zinc protoporphyrin. *Clin Lab Haematol* 12:169-176.
364. Monti, D., B. Vodopivec, N. Basilico, P. Olliaro, and D. Taramelli. 1999. A novel endogenous antimalarial: Fe(II)-protoporphyrin IX alpha (heme) inhibits hematin polymerization to beta-hematin (malaria pigment) and kills malaria parasites. *Biochemistry* 38:8858-8863.
365. O'Neill, P. M., L. P. Bishop, R. C. Storr, S. R. Hawley, J. L. Maggs, S. A. Ward, and B. K. Park. 1996. Mechanism-based design of parasite-targeted artemisinin derivatives: synthesis and antimalarial activity of benzylamino and alkylamino ether analogues of artemisinin. *J Med Chem* 39:4511-4514.
366. Becuwe, P., S. Gratepanche, M. N. Fourmaux, J. Van Beeumen, B. Samyn, O. Mercereau-Puijalon, J. P. Touzel, C. Slomianny, D. Camus, and D. Dive. 1996. Characterization of iron-dependent endogenous superoxide dismutase of Plasmodium falciparum. *Mol Biochem Parasitol* 76:125-134.
367. Gordeuk, V. R., P. E. Thuma, and G. M. Brittenham. 1994. Iron chelation therapy for malaria. *Adv Exp Med Biol* 356:371-383.
368. Whitehead, S., and T. E. Peto. 1990. Stage-dependent effect of deferoxamine on growth of Plasmodium falciparum in vitro. *Blood* 76:1250-1255.
369. Van Zyl, R. L., I. Havlik, E. Hempelmann, A. P. MacPhail, and L. McNamara. 1993. Malaria pigment and extracellular iron. Possible target for iron chelating agents. *Biochem Pharmacol* 45:1431-1436.
370. Aruoma, O. I., B. Halliwell, E. Gajewski, and M. Dizdaroglu. 1989. Damage to the bases in DNA induced by hydrogen peroxide and ferric ion chelates. *J Biol Chem* 264:20509-20512.
371. Van Zyl, R. L., I. Havlik, and F. S. Monteagudo. 1992. The combined effect of iron chelators and classical antimalarials on the in-vitro growth of Plasmodium falciparum. *J Antimicrob Chemother* 30:273-278.
372. Tiffert, T., H. Ginsburg, M. Krugliak, B. C. Elford, and V. L. Lew. 2000. Potent antimalarial activity of clotrimazole in in vitro cultures of Plasmodium falciparum. *Proc Natl Acad Sci U S A* 97:331-336.
373. Huy, N. T., K. Kamei, T. Yamamoto, Y. Kondo, K. Kanaori, R. Takano, K. Tajima, and S. Hara. 2002. Clotrimazole binds to heme and enhances heme-dependent hemolysis: proposed antimalarial mechanism of clotrimazole. *J Biol Chem* 277:4152-4158.

374. Winter, R. W., M. Ignatushchenko, O. A. Ogundahunsi, K. A. Cornell, A. M. Oduola, D. J. Hinrichs, and M. K. Riscoe. 1997. Potentiation of an antimalarial oxidant drug. *Antimicrob Agents Chemother* 41:1449-1454.
375. Ignatushchenko, M. V., R. W. Winter, H. P. Bachinger, D. J. Hinrichs, and M. K. Riscoe. 1997. Xanthones as antimalarial agents; studies of a possible mode of action. *FEBS Lett* 409:67-73.
376. Calas, M., M. L. Ancelin, G. Cordina, P. Portefaix, G. Piquet, V. Vidal-Sailhan, and H. Vial. 2000. Antimalarial activity of compounds interfering with Plasmodium falciparum phospholipid metabolism: comparison between mono- and bisquaternary ammonium salts. *J Med Chem* 43:505-516.
377. Wengelnik, K., V. Vidal, M. L. Ancelin, A. M. Cathiard, J. L. Morgat, C. H. Kocken, M. Calas, S. Herrera, A. W. Thomas, and H. J. Vial. 2002. A class of potent antimalarials and their specific accumulation in infected erythrocytes. *Science* 295:1311-1314.
378. Biagini, G. A., E. Richier, P. G. Bray, M. Calas, H. Vial, and S. A. Ward. 2003. Heme binding contributes to antimalarial activity of bis-quaternary ammoniums. *Antimicrob Agents Chemother* 47:2584-2589.
379. Hyde, J. E. 2002. Mechanisms of resistance of Plasmodium falciparum to antimalarial drugs. *Microbes Infect* 4:165-174.
380. Auparakkitanon, S., and P. Wilairat. 2000. Cleavage of DNA induced by 9-anilinoacridine inhibitors of topoisomerase II in the malaria parasite Plasmodium falciparum. *Biochem Biophys Res Commun* 269:406-409.
381. Auparakkitanon, S., W. Noonpakdee, R. K. Ralph, W. A. Denny, and P. Wilairat. 2003. Antimalarial 9-anilinoacridine compounds directed at hematin. *Antimicrob Agents Chemother* 47:3708-3712.
382. Vennerstrom, J. L., M. T. Makler, C. K. Angerhofer, and J. A. Williams. 1995. Antimalarial dyes revisited: xanthenes, azines, oxazines, and thiazines. *Antimicrob Agents Chemother* 39:2671-2677.
383. Atamna, H., M. Krugliak, G. Shalmiev, E. Deharo, G. Pescarmona, and H. Ginsburg. 1996. Mode of antimalarial effect of methylene blue and some of its analogues on Plasmodium falciparum in culture and their inhibition of P. vinckei petteri and P. yoelii nigeriensis in vivo. *Biochem Pharmacol* 51:693-700.
384. Shear, H. L., R. S. Nussenzweig, and C. Bianco. 1979. Immune phagocytosis in murine malaria. *J Exp Med* 149:1288-1298.
385. Celada, A., A. Cruchaud, and L. H. Perrin. 1983. Phagocytosis of Plasmodium falciparum-parasitized erythrocytes by human polymorphonuclear leukocytes. *J Parasitol* 69:49-53.
386. Nguyen, P. H., N. Day, T. D. Pram, D. J. Ferguson, and N. J. White. 1995. Intraleucocytic malaria pigment and prognosis in severe malaria. *Trans R Soc Trop Med Hyg* 89:200-204.
387. Metzger, W. G., B. G. Mordmuller, and P. G. Kremsner. 1995. Malaria pigment in leucocytes. *Trans R Soc Trop Med Hyg* 89:637-638.
388. Amodu, O. K., A. A. Adeyemo, P. E. Olumese, and R. A. Gbadegesin. 1998. Intraleucocytic malaria pigment and clinical severity of malaria in children. *Trans R Soc Trop Med Hyg* 92:54-56.
389. Schwarzer, E., F. De Matteis, G. Giribaldi, D. Ulliers, E. Valente, and P. Arese. 1999. Hemozoin stability and dormant induction of heme oxygenase in hemozoin-fed human monocytes. *Mol Biochem Parasitol* 100:61-72.
390. Lawrence, C., and J. A. Olson. 1986. Birefringent hemozoin identifies malaria. *Am J Clin Pathol* 86:360-363.
391. Day, N. P., T. D. Pham, T. L. Phan, X. S. Dinh, P. L. Pham, V. C. Ly, T. H. Tran, T. H. Nguyen, D. B. Bethell, H. P. Nguyan, and N. J. White. 1996. Clearance kinetics of parasites and pigment-containing leukocytes in severe malaria. *Blood* 88:4694-4700.
392. Lawrence, C. 1999. Laveran remembered: malaria haemozoin in leucocytes. *Lancet* 353:1852.
393. Edington, G. M. 1967. Pathology of malaria in West Africa. *Br Med J* 1:715-718.
394. Taliaferro, W. H., and H. W. Mulligan. 1937. The histopathology of malaria with special reference to the function and origin of macrophages in defense. *Indian Med. Mem.* 29:1.
395. Newton, C. R., T. E. Taylor, and R. O. Whitten. 1998. Pathophysiology of fatal falciparum malaria in African children. *Am J Trop Med Hyg* 58:673-683.
396. Aikawa, M., M. Suzuki, and Y. Guttierrez. 1980. Pathology of Malaria. In *Malaria*. J. P. Kreier, ed. Academic Press, New York.
397. Scheibel, L. W., and I. W. Sherman. 1988. Plasmodial metabolism and related organellar function during various stages of the life-cylcle proteins, lipids, nucleic acids and vitamins. In *Malaria: Principles and Practice of Malariology*. W. H. Wernsdorfer, and I. McGregor, eds. Churchill-Livingstone, Edinburgh.
398. Levesque, M. A., A. D. Sullivan, and S. R. Meshnick. 1999. Splenic and hepatic hemozoin in mice after malaria parasite clearance. *J Parasitol* 85:570-573.
399. Sullivan, A. D., I. Ittarat, and S. R. Meshnick. 1996. Patterns of haemozoin accumulation in tissue. *Parasitology* 112 (Pt 3):285-294.
400. Schwarzer, E., G. Bellomo, G. Giribaldi, D. Ulliers, and P. Arese. 2001. Phagocytosis of malarial pigment haemozoin by human monocytes: a confocal microscopy study. *Parasitology* 123:125-131.
401. Fiori, P. L., P. Rappelli, S. N. Mirkarimi, H. Ginsburg, P. Cappuccinelli, and F. Turrini. 1993. Reduced microbicidal and anti-tumour activities of human monocytes after ingestion of Plasmodium falciparum-infected red blood cells. *Parasite Immunol* 15:647-655.
402. Gopalakrishna, R., and W. B. Anderson. 1989. Ca2+- and phospholipid-independent activation of protein kinase C by selective oxidative modification of the regulatory domain. *Proc Natl Acad Sci U S A* 86:6758-6762.

403. Bussolino, F., E. Fischer, F. Turrini, M. D. Kazatchkine, and P. Arese. 1989. Platelet-activating factor enhances complement-dependent phagocytosis of diamide-treated erythrocytes by human monocytes through activation of protein kinase C and phosphorylation of complement receptor type one (CR1). *J Biol Chem* 264:21711-21719.

404. Nauseef, W. M., B. D. Volpp, S. McCormick, K. G. Leidal, and R. A. Clark. 1991. Assembly of the neutrophil respiratory burst oxidase. Protein kinase C promotes cytoskeletal and membrane association of cytosolic oxidase components. *J Biol Chem* 266:5911-5917.

405. Hubbard, S. R., W. R. Bishop, P. Kirschmeier, S. J. George, S. P. Cramer, and W. A. Hendrickson. 1991. Identification and characterization of zinc binding sites in protein kinase C. *Science* 254:1776-1779.

406. Segal, A. W., and A. Abo. 1993. The biochemical basis of the NADPH oxidase of phagocytes. *Trends Biochem Sci* 18:43-47.

407. Esterbauer, H., R. J. Schaur, and H. Zollner. 1991. Chemistry and biochemistry of 4-hydroxynonenal, malonaldehyde and related aldehydes. *Free Radic Biol Med* 11:81-128.

408. Uchida, K., and E. R. Stadtman. 1992. Modification of histidine residues in proteins by reaction with 4-hydroxynonenal. *Proc Natl Acad Sci U S A* 89:4544-4548.

409. Green, M. D., L. Xiao, and A. A. Lal. 1996. Formation of hydroxyeicosatetraenoic acids from hemozoin-catalyzed oxidation of arachidonic acid. *Mol Biochem Parasitol* 83:183-188.

410. Takahashi, M., N. Ozaki, S. Kawakita, M. Nozaki, and Y. Saeki. 1985. Vascular effects of 15-hydroperoxyeicosatetraenoic acid and 15-hydroxyeicosatetraenoic acid on canine arteries. *Jpn J Pharmacol* 37:325-334.

411. Villacara, A., M. Spatz, R. F. Dodson, C. Corn, and J. Bembry. 1989. Effect of arachidonic acid on cultured cerebromicrovascular endothelium: permeability, lipid peroxidation and membrane "fluidity". *Acta Neuropathol* 78:310-316.

412. Setty, B. N., C. D. Dampier, and M. J. Stuart. 1995. Arachidonic acid metabolites are involved in mediating red blood cell adherence to endothelium. *J Lab Clin Med* 125:608-617.

413. Kuitert, L. M., R. Newton, N. C. Barnes, I. M. Adcock, and P. J. Barnes. 1996. Eicosanoid mediator expression in mononuclear and polymorphonuclear cells in normal subjects and patients with atopic asthma and cystic fibrosis. *Thorax* 51:1223-1228.

414. Schwarzer, E., O. Muller, P. Arese, W. G. Siems, and T. Grune. 1996. Increased levels of 4-hydroxynonenal in human monocytes fed with malarial pigment hemozoin. A possible clue for hemozoin toxicity. *FEBS Lett* 388:119-122.

415. Perkins, D. J., P. G. Kremsner, and J. B. Weinberg. 2001. Inverse relationship of plasma prostaglandin E2 and blood mononuclear cell cyclooxygenase-2 with disease severity in children with Plasmodium falciparum malaria. *J Infect Dis* 183:113-118.

416. Gordon, E. E., J. A. Gordon, and A. A. Spector. 1991. HETEs and coronary artery endothelial cells: metabolic and functional interactions. *Am J Physiol* 261:C623-633.

417. Duval, D. L., D. J. Sieg, and R. E. Billings. 1995. Regulation of hepatic nitric oxide synthase by reactive oxygen intermediates and glutathione. *Arch Biochem Biophys* 316:699-706.

418. Jaimes, E. A., K. A. Nath, and L. Raij. 1997. Hydrogen peroxide downregulates IL-1-driven mesangial iNOS activity: implications for glomerulonephritis. *Am J Physiol* 272:F721-728.

419. Wong, H. R., M. Ryan, and J. R. Wispe. 1997. Stress response decreases NF-kappaB nuclear translocation and increases I-kappaBalpha expression in A549 cells. *J Clin Invest* 99:2423-2428.

420. Poss, K. D., and S. Tonegawa. 1997. Reduced stress defense in heme oxygenase 1-deficient cells. *Proc Natl Acad Sci U S A* 94:10925-10930.

421. Omodeo-Sale, F., N. Basilico, M. Folini, P. Olliaro, and D. Taramelli. 1998. Macrophage populations of different origins have distinct susceptibilities to lipid peroxidation induced by beta-haematin (malaria pigment). *FEBS Lett* 433:215-218.

422. Boss, J. M. 1997. Regulation of transcription of MHC class II genes. *Curr Opin Immunol* 9:107-113.

423. Fan, X. D., M. Goldberg, and B. R. Bloom. 1988. Interferon-gamma-induced transcriptional activation is mediated by protein kinase C. *Proc Natl Acad Sci U S A* 85:5122-5125.

424. Blackford, J., H. W. Reid, D. J. Pappin, F. S. Bowers, and J. M. Wilkinson. 1996. A monoclonal antibody, 3/22, to rabbit CD11c which induces homotypic T cell aggregation: evidence that ICAM-1 is a ligand for CD11c/CD18. *Eur J Immunol* 26:525-531.

425. Loose, L. D. 1984. Characterization of macrophage dysfunction in rodent malaria. *J Leukoc Biol* 36:703-718.

426. Clark, I. A., F. M. al Yaman, and L. S. Jacobson. 1997. The biological basis of malarial disease. *Int J Parasitol* 27:1237-1249.

427. Arese, P., and E. Schwarzer. 1997. Malarial pigment (haemozoin): a very active 'inert' substance. *Ann Trop Med Parasitol* 91:501-516.

428. Kremsner, P. G., P. Radloff, W. Metzger, E. Wildling, B. Mordmuller, J. Philipps, L. Jenne, M. Nkeyi, J. Prada, U. Bienzle, and et al. 1995. Quinine plus clindamycin improves chemotherapy of severe malaria in children. *Antimicrob Agents Chemother* 39:1603-1605.

429. Mohan, K., and M. M. Stevenson. 1998. Interleukin-12 corrects severe anemia during blood-stage Plasmodium chabaudi AS in susceptible A/J mice. *Exp Hematol* 26:45-52.

430. Metz, C. N., and R. Bucala. 1997. Role of macrophage migration inhibitory factor in the regulation of the immune response. *Adv Immunol* 66:197-223.

431. Abdalla, S. H. 1988. Peripheral blood and bone marrow leucocytes in Gambian children with malaria: numerical changes and evaluation of phagocytosis. *Ann Trop Paediatr* 8:250-258.

432. Kishimoto, T. K., R. A. Warnock, M. A. Jutila, E. C. Butcher, C. Lane, D. C. Anderson, and C. W. Smith. 1991. Antibodies against human neutrophil LECAM-1 (LAM-1/Leu-8/DREG-56 antigen) and endothelial cell ELAM-1 inhibit a common CD18-independent adhesion pathway in vitro. *Blood* 78:805-811.

433. Kishimoto, T. K., M. A. Jutila, E. L. Berg, and E. C. Butcher. 1989. Neutrophil Mac-1 and MEL-14 adhesion proteins inversely regulated by chemotactic factors. *Science* 245:1238-1241.

434. Griffin, J. D., O. Spertini, T. J. Ernst, M. P. Belvin, H. B. Levine, Y. Kanakura, and T. F. Tedder. 1990. Granulocyte-macrophage colony-stimulating factor and other cytokines regulate surface expression of the leukocyte adhesion molecule-1 on human neutrophils, monocytes, and their precursors. *J Immunol* 145:576-584.

435. Mellman, I., and R. M. Steinman. 2001. Dendritic cells: specialized and regulated antigen processing machines. *Cell* 106:255-258.

436. Reis e Sousa, C. 2001. Dendritic cells as sensors of infection. *Immunity* 14:495-498.

437. Takashima, A., and A. Morita. 1999. Dendritic cells in genetic immunization. *J Leukoc Biol* 66:350-356.

438. MacMicking, J., Q. W. Xie, and C. Nathan. 1997. Nitric oxide and macrophage function. *Annu Rev Immunol* 15:323-350.

439. Palmer, R. M., A. G. Ferrige, and S. Moncada. 1987. Nitric oxide release accounts for the biological activity of endothelium-derived relaxing factor. *Nature* 327:524-526.

440. Garthwaite, J. 1991. Glutamate, nitric oxide and cell-cell signalling in the nervous system. *Trends Neurosci* 14:60-67.

441. James, S. L. 1995. Role of nitric oxide in parasitic infections. *Microbiol Rev* 59:533-547.

442. Bogdan, C. 2001. Nitric oxide and the immune response. *Nat Immunol* 2:907-916.

443. Culotta, E., and D. E. Koshland, Jr. 1992. NO news is good news. *Science* 258:1862-1865.

444. Marletta, M. A. 1994. Nitric oxide synthase: aspects concerning structure and catalysis. *Cell* 78:927-930.

445. Bredt, D. S., P. M. Hwang, C. E. Glatt, C. Lowenstein, R. R. Reed, and S. H. Snyder. 1991. Cloned and expressed nitric oxide synthase structurally resembles cytochrome P-450 reductase. *Nature* 351:714-718.

446. Pollock, J. S., U. Forstermann, J. A. Mitchell, T. D. Warner, H. H. Schmidt, M. Nakane, and F. Murad. 1991. Purification and characterization of particulate endothelium-derived relaxing factor synthase from cultured and native bovine aortic endothelial cells. *Proc Natl Acad Sci U S A* 88:10480-10484.

447. Bredt, D. S., and S. H. Snyder. 1990. Isolation of nitric oxide synthetase, a calmodulin-requiring enzyme. *Proc Natl Acad Sci U S A* 87:682-685.

448. Ding, A. H., C. F. Nathan, and D. J. Stuehr. 1988. Release of reactive nitrogen intermediates and reactive oxygen intermediates from mouse peritoneal macrophages. Comparison of activating cytokines and evidence for independent production. *J Immunol* 141:2407-2412.

449. Lowenstein, C. J., C. S. Glatt, D. S. Bredt, and S. H. Snyder. 1992. Cloned and expressed macrophage nitric oxide synthase contrasts with the brain enzyme. *Proc Natl Acad Sci U S A* 89:6711-6715.

450. Xie, Q. W., H. J. Cho, J. Calaycay, R. A. Mumford, K. M. Swiderek, T. D. Lee, A. Ding, T. Troso, and C. Nathan. 1992. Cloning and characterization of inducible nitric oxide synthase from mouse macrophages. *Science* 256:225-228.

451. Nathan, C., and Q. W. Xie. 1994. Nitric oxide synthases: roles, tolls, and controls. *Cell* 78:915-918.

452. Matte, C., J. F. Marquis, J. Blanchette, P. Gros, R. Faure, B. I. Posner, and M. Olivier. 2000. Peroxovanadium-mediated protection against murine leishmaniasis: role of the modulation of nitric oxide. *Eur J Immunol* 30:2555-2564.

453. Cook, H. T., A. Jansen, S. Lewis, P. Largen, M. O'Donnell, D. Reaveley, and V. Cattell. 1994. Arginine metabolism in experimental glomerulonephritis: interaction between nitric oxide synthase and arginase. *Am J Physiol* 267:F646-653.

454. Stuehr, D. J., O. A. Fasehun, N. S. Kwon, S. S. Gross, J. A. Gonzalez, R. Levi, and C. F. Nathan. 1991. Inhibition of macrophage and endothelial cell nitric oxide synthase by diphenyleneiodonium and its analogs. *FASEB J* 5:98-103.

455. Griscavage, J. M., N. E. Rogers, M. P. Sherman, and L. J. Ignarro. 1993. Inducible nitric oxide synthase from a rat alveolar macrophage cell line is inhibited by nitric oxide. *J Immunol* 151:6329-6337.

456. Hibbs, J. B., Jr., R. R. Taintor, Z. Vavrin, and E. M. Rachlin. 1988. Nitric oxide: a cytotoxic activated macrophage effector molecule. *Biochem Biophys Res Commun* 157:87-94.

457. Chen, B., M. Keshive, and W. M. Deen. 1998. Diffusion and reaction of nitric oxide in suspension cell cultures. *Biophys J* 75:745-754.

458. Woods, M. L., 2nd, J. Mayer, T. G. Evans, and J. B. Hibbs, Jr. 1994. Antiparasitic effects of nitric oxide in an in vitro murine model of Chlamydia trachomatis infection and an in vivo murine model of Leishmania major infection. *Immunol Ser* 60:179-195.
459. Beckman, J. S., T. W. Beckman, J. Chen, P. A. Marshall, and B. A. Freeman. 1990. Apparent hydroxyl radical production by peroxynitrite: implications for endothelial injury from nitric oxide and superoxide. *Proc Natl Acad Sci U S A* 87:1620-1624.
460. Dedon, P. C., and S. R. Tannenbaum. 2004. Reactive nitrogen species in the chemical biology of inflammation. *Arch Biochem Biophys* 423:12-22.
461. Stamler, J. S., D. J. Singel, and J. Loscalzo. 1992. Biochemistry of nitric oxide and its redox-activated forms. *Science* 258:1898-1902.
462. Kwon, N. S., D. J. Stuehr, and C. F. Nathan. 1991. Inhibition of tumor cell ribonucleotide reductase by macrophage-derived nitric oxide. *J Exp Med* 174:761-767.
463. Lancaster, J. R., Jr., and J. B. Hibbs, Jr. 1990. EPR demonstration of iron-nitrosyl complex formation by cytotoxic activated macrophages. *Proc Natl Acad Sci U S A* 87:1223 1227.
464. van der Vliet, A., J. P. Eiserich, B. Halliwell, and C. E. Cross. 1997. Formation of reactive nitrogen species during peroxidase-catalyzed oxidation of nitrite. A potential additional mechanism of nitric oxide-dependent toxicity. *J Biol Chem* 272:7617-7625.
465. Dawson, T. M., V. L. Dawson, and S. H. Snyder. 1992. A novel neuronal messenger molecule in brain: the free radical, nitric oxide. *Ann Neurol* 32:297-311.
466. Marletta, M. A. 1989. Nitric oxide: biosynthesis and biological significance. *Trends Biochem Sci* 14:488-492.
467. Moncada, S., R. M. Palmer, and E. A. Higgs. 1991. Nitric oxide: physiology, pathophysiology, and pharmacology. *Pharmacol Rev* 43:109-142.
468. Bobe, P., K. Benihoud, D. Grandjon, P. Opolon, L. L. Pritchard, and R. Huchet. 1999. Nitric oxide mediation of active immunosuppression associated with graft-versus-host reaction. *Blood* 94:1028-1037.
469. Shi, F. D., M. Flodstrom, S. H. Kim, S. Pakala, M. Cleary, H. G. Ljunggren, and N. Sarvetnick. 2001. Control of the autoimmune response by type 2 nitric oxide synthase. *J Immunol* 167:3000-3006.
470. Schindler, H., and C. Bogdan. 2001. NO as a signaling molecule: effects on kinases. *Int Immunopharmacol* 1:1443-1455.
471. Cherla, R. P., and R. K. Ganju. 2001. Stromal cell-derived factor 1 alpha-induced chemotaxis in T cells is mediated by nitric oxide signaling pathways. *J Immunol* 166:3067-3074.
472. Williams, M. S., S. Noguchi, P. A. Henkart, and Y. Osawa. 1998. Nitric oxide synthase plays a signaling role in TCR-triggered apoptotic death. *J Immunol* 161:6526-6531.
473. Fang, F. C. 1997. Perspectives series: host/pathogen interactions. Mechanisms of nitric oxide-related antimicrobial activity. *J Clin Invest* 99:2818-2825.
474. Evans, T. G., L. Thai, D. L. Granger, and J. B. Hibbs, Jr. 1993. Effect of in vivo inhibition of nitric oxide production in murine leishmaniasis. *J Immunol* 151:907-915.
475. Nicholson, S., G. Bonecini-Almeida Mda, J. R. Lapa e Silva, C. Nathan, Q. W. Xie, R. Mumford, J. R. Weidner, J. Calaycay, J. Geng, N. Boechat, C. Linhares, W. Rom, and J. L. Ho. 1996. Inducible nitric oxide synthase in pulmonary alveolar macrophages from patients with tuberculosis. *J Exp Med* 183:2293-2302.
476. Stenger, S., N. Donhauser, H. Thuring, M. Rollinghoff, and C. Bogdan. 1996. Reactivation of latent leishmaniasis by inhibition of inducible nitric oxide synthase. *J Exp Med* 183:1501-1514.
477. MacMicking, J. D., R. J. North, R. LaCourse, J. S. Mudgett, S. K. Shah, and C. F. Nathan. 1997. Identification of nitric oxide synthase as a protective locus against tuberculosis. *Proc Natl Acad Sci U S A* 94:5243-5248.
478. Ahmed, S. F., I. P. Oswald, P. Caspar, S. Hieny, L. Keefer, A. Sher, and S. L. James. 1997. Developmental differences determine larval susceptibility to nitric oxide-mediated killing in a murine model of vaccination against Schistosoma mansoni. *Infect Immun* 65:219-226.
479. Marletta, M. A., P. S. Yoon, R. Iyengar, C. D. Leaf, and J. S. Wishnok. 1988. Macrophage oxidation of L-arginine to nitrite and nitrate: nitric oxide is an intermediate. *Biochemistry* 27:8706-8711.
480. Hibbs, J. B., Jr., Z. Vavrin, and R. R. Taintor. 1987. L-arginine is required for expression of the activated macrophage effector mechanism causing selective metabolic inhibition in target cells. *J Immunol* 138:550-565.
481. Murray, H. W., and R. F. Teitelbaum. 1992. L-arginine-dependent reactive nitrogen intermediates and the antimicrobial effect of activated human mononuclear phagocytes. *J Infect Dis* 165:513-517.
482. Schneemann, M., G. Schoedon, S. Hofer, N. Blau, L. Guerrero, and A. Schaffner. 1993. Nitric oxide synthase is not a constituent of the antimicrobial armature of human mononuclear phagocytes. *J Infect Dis* 167:1358-1363.
483. Bukrinsky, M. I., H. S. Nottet, H. Schmidtmayerova, L. Dubrovsky, C. R. Flanagan, M. E. Mullins, S. A. Lipton, and H. E. Gendelman. 1995. Regulation of nitric oxide synthase activity in human immunodeficiency virus type 1 (HIV-1)-infected monocytes: implications for HIV-associated neurological disease. *J Exp Med* 181:735-745.
484. De Maria, R., M. G. Cifone, R. Trotta, M. R. Rippo, C. Festuccia, A. Santoni, and R. Testi. 1994. Triggering of human monocyte activation through CD69, a member of the natural killer cell gene complex family of signal transducing receptors. *J Exp Med* 180:1999-2004.

485. Vouldoukis, I., V. Riveros-Moreno, B. Dugas, F. Ouaaz, P. Becherel, P. Debre, S. Moncada, and M. D. Mossalayi. 1995. The killing of Leishmania major by human macrophages is mediated by nitric oxide induced after ligation of the Fc epsilon RII/CD23 surface antigen. *Proc Natl Acad Sci U S A* 92:7804-7808.

486. Adamson, D. C., B. Wildemann, M. Sasaki, J. D. Glass, J. C. McArthur, V. I. Christov, T. M. Dawson, and V. L. Dawson. 1996. Immunologic NO synthase: elevation in severe AIDS dementia and induction by HIV-1 gp41. *Science* 274:1917-1921.

487. Kroncke, K. D., K. Fehsel, and V. Kolb-Bachofen. 1998. Inducible nitric oxide synthase in human diseases. *Clin Exp Immunol* 113:147-156.

488. St Clair, E. W., W. E. Wilkinson, T. Lang, L. Sanders, M. A. Misukonis, G. S. Gilkeson, D. S. Pisetsky, D. I. Granger, and J. B. Weinberg. 1996. Increased expression of blood mononuclear cell nitric oxide synthase type 2 in rheumatoid arthritis patients. *J Exp Med* 184:1173-1178.

489. Weinberg, J. B. 1998. Nitric oxide production and nitric oxide synthase type 2 expression by human mononuclear phagocytes: a review. *Mol Med* 4:557-591.

490. Gyan, B., M. Troye-Blomberg, P. Perlmann, and A. Bjorkman. 1994. Human monocytes cultured with and without interferon-gamma inhibit Plasmodium falciparum parasite growth in vitro via secretion of reactive nitrogen intermediates. *Parasite Immunol* 16:371-375.

491. Munoz-Fernandez, M. A., M. A. Fernandez, and M. Fresno. 1992. Activation of human macrophages for the killing of intracellular Trypanosoma cruzi by TNF-alpha and IFN-gamma through a nitric oxide-dependent mechanism. *Immunol Lett* 33:35-40.

492. Jimenez, J. L., J. Gonzalez-Nicolas, S. Alvarez, M. Fresno, and M. A. Munoz-Fernandez. 2001. Regulation of human immunodeficiency virus type 1 replication in human T lymphocytes by nitric oxide. *J Virol* 75:4655-4663.

493. Nabeshima, S., M. Nomoto, G. Matsuzaki, K. Kishihara, H. Taniguchi, S. Yoshida, and K. Nomoto. 1999. T-Cell hyporesponsiveness induced by activated macrophages through nitric oxide production in mice infected with Mycobacterium tuberculosis. *Infect Immun* 67:3221-3226.

494. Clancy, R. M., A. R. Amin, and S. B. Abramson. 1998. The role of nitric oxide in inflammation and immunity. *Arthritis Rheum* 41:1141-1151.

495. Saeftel, M., B. Fleischer, and A. Hoerauf. 2001. Stage-dependent role of nitric oxide in control of Trypanosoma cruzi infection. *Infect Immun* 69:2252-2259.

496. Kim, Y. M., H. Bergonia, and J. R. Lancaster, Jr. 1995. Nitrogen oxide-induced autoprotection in isolated rat hepatocytes. *FEBS Lett* 374:228-232.

497. Unno, N., H. Wang, M. J. Menconi, S. H. Tytgat, V. Larkin, M. Smith, M. J. Morin, A. Chavez, R. A. Hodin, and M. P. Fink. 1997. Inhibition of inducible nitric oxide synthase ameliorates endotoxin-induced gut mucosal barrier dysfunction in rats. *Gastroenterology* 113:1246-1257.

498. Brunet, L. R. 2001. Nitric oxide in parasitic infections. *Int Immunopharmacol* 1:1457-1467.

499. Rodriguez-Pascual, F., M. Hausding, I. Ihrig-Biedert, H. Furneaux, A. P. Levy, U. Forstermann, and H. Kleinert. 2000. Complex contribution of the 3'-untranslated region to the expressional regulation of the human inducible nitric-oxide synthase gene. Involvement of the RNA-binding protein HuR. *J Biol Chem* 275:26040-26049.

500. Fritsche, G., C. Larcher, H. Schennach, and G. Weiss. 2001. Regulatory interactions between iron and nitric oxide metabolism for immune defense against Plasmodium falciparum infection. *J Infect Dis* 183:1388-1394.

501. Frucht, D. M., T. Fukao, C. Bogdan, H. Schindler, J. J. O'Shea, and S. Koyasu. 2001. IFN-gamma production by antigen-presenting cells: mechanisms emerge. *Trends Immunol* 22:556-560.

502. Gao, J. J., E. G. Zuvanich, Q. Xue, D. L. Horn, R. Silverstein, and D. C. Morrison. 1999. Cutting edge: bacterial DNA and LPS act in synergy in inducing nitric oxide production in RAW 264.7 macrophages. *J Immunol* 163:4095-4099.

503. Mori, N., Y. Nunokawa, Y. Yamada, S. Ikeda, M. Tomonaga, and N. Yamamoto. 1999. Expression of human inducible nitric oxide synthase gene in T-cell lines infected with human T-cell leukemia virus type-I and primary adult T-cell leukemia cells. *Blood* 94:2862-2870.

504. Shoda, L. K., K. A. Kegerreis, C. E. Suarez, I. Roditi, R. S. Corral, G. M. Bertot, J. Norimine, and W. C. Brown. 2001. DNA from protozoan parasites Babesia bovis, Trypanosoma cruzi, and T. brucei is mitogenic for B lymphocytes and stimulates macrophage expression of interleukin-12, tumor necrosis factor alpha, and nitric oxide. *Infect Immun* 69:2162-2171.

505. Thoma-Uszynski, S., S. Stenger, O. Takeuchi, M. T. Ochoa, M. Engele, P. A. Sieling, P. F. Barnes, M. Rollinghoff, P. L. Bolcskei, M. Wagner, S. Akira, M. V. Norgard, J. T. Belisle, P. J. Godowski, B. R. Bloom, and R. L. Modlin. 2001. Induction of direct antimicrobial activity through mammalian toll-like receptors. *Science* 291:1544-1547.

506. Dlaska, M., and G. Weiss. 1999. Central role of transcription factor NF-IL6 for cytokine and iron-mediated regulation of murine inducible nitric oxide synthase expression. *J Immunol* 162:6171-6177.

507. Ganster, R. W., B. S. Taylor, L. Shao, and D. A. Geller. 2001. Complex regulation of human inducible nitric oxide synthase gene transcription by Stat 1 and NF-kappa B. *Proc Natl Acad Sci U S A* 98:8638-8643.

508. Kleinert, H., T. Wallerath, G. Fritz, I. Ihrig-Biedert, F. Rodriguez-Pascual, D. A. Geller, and U. Forstermann. 1998. Cytokine induction of NO synthase II in human DLD-1 cells: roles of the JAK-STAT, AP-1 and NF-kappaB-signaling pathways. *Br J Pharmacol* 125:193-201.

509. Blanchette, J., M. Jaramillo, and M. Olivier. 2003. Signalling events involved in interferon-gamma-inducible macrophage nitric oxide generation. *Immunology* 108:513-522.

510. Chakravortty, D., Y. Kato, T. Sugiyama, N. Koide, M. M. Mu, T. Yoshida, and T. Yokochi. 2001. The inhibitory action of sodium arsenite on lipopolysaccharide-induced nitric oxide production in RAW 267.4 macrophage cells: a role of Raf-1 in lipopolysaccharide signaling. *J Immunol* 166:2011-2017.

511. Chan, E. D., K. R. Morris, J. T. Belisle, P. Hill, L. K. Remigio, P. J. Brennan, and D. W. Riches. 2001. Induction of inducible nitric oxide synthase-NO* by lipoarabinomannan of Mycobacterium tuberculosis is mediated by MEK1-ERK, MKK7-JNK, and NF-kappaB signaling pathways. *Infect Immun* 69:2001-2010.

512. Karaghiosoff, M., H. Neubauer, C. Lassnig, P. Kovarik, H. Schindler, H. Pircher, B. McCoy, C. Bogdan, T. Decker, G. Brem, K. Pfeffer, and M. Muller. 2000. Partial impairment of cytokine responses in Tyk2-deficient mice. *Immunity* 13:549-560.

513. Mullet, D., R. H. Fertel, D. Kniss, and G. W. Cox. 1997. An increase in intracellular cyclic AMP modulates nitric oxide production in IFN-gamma-treated macrophages. *J Immunol* 158:897-904.

514. Castrillo, A., D. J. Pennington, F. Otto, P. J. Parker, M. J. Owen, and L. Bosca. 2001. Protein kinase Cepsilon is required for macrophage activation and defense against bacterial infection. *J Exp Med* 194:1231-1242.

515. Dong, Z., X. Yang, K. Xie, S. H. Juang, N. Llansa, and I. J. Fidler. 1995. Activation of inducible nitric oxide synthase gene in murine macrophages requires protein phosphatases 1 and 2A activities. *J Leukoc Biol* 58:725-732.

516. Kristof, A. S., J. Marks-Konczalik, and J. Moss. 2001. Mitogen-activated protein kinases mediate activator protein-1-dependent human inducible nitric-oxide synthase promoter activation. *J Biol Chem* 276:8445-8452.

517. Diaz-Guerra, M. J., A. Castrillo, P. Martin-Sanz, and L. Bosca. 1999. Negative regulation by phosphatidylinositol 3-kinase of inducible nitric oxide synthase expression in macrophages. *J Immunol* 162:6184-6190.

518. Pahan, K., X. Liu, C. Wood, and J. R. Raymond. 2000. Expression of a constitutively active form of phosphatidylinositol 3-kinase inhibits the induction of nitric oxide synthase in human astrocytes. *FEBS Lett* 472:203-207.

519. Umansky, V., S. P. Hehner, A. Dumont, T. G. Hofmann, V. Schirrmacher, W. Droge, and M. L. Schmitz. 1998. Co-stimulatory effect of nitric oxide on endothelial NF-kappaB implies a physiological self-amplifying mechanism. *Eur J Immunol* 28:2276-2282.

520. Connelly, L., M. Palacios-Callender, C. Ameixa, S. Moncada, and A. J. Hobbs. 2001. Biphasic regulation of NF-kappa B activity underlies the pro- and anti-inflammatory actions of nitric oxide. *J Immunol* 166:3873-3881.

521. Xie, Q. W., R. Whisnant, and C. Nathan. 1993. Promoter of the mouse gene encoding calcium-independent nitric oxide synthase confers inducibility by interferon gamma and bacterial lipopolysaccharide. *J Exp Med* 177:1779-1784.

522. Xie, Q. W., Y. Kashiwabara, and C. Nathan. 1994. Role of transcription factor NF-kappa B/Rel in induction of nitric oxide synthase. *J Biol Chem* 269:4705-4708.

523. Griscavage, J. M., S. Wilk, and L. J. Ignarro. 1996. Inhibitors of the proteasome pathway interfere with induction of nitric oxide synthase in macrophages by blocking activation of transcription factor NF-kappa B. *Proc Natl Acad Sci U S A* 93:3308-3312.

524. Kim, Y. M., B. S. Lee, K. Y. Yi, and S. G. Paik. 1997. Upstream NF-kappaB site is required for the maximal expression of mouse inducible nitric oxide synthase gene in interferon-gamma plus lipopolysaccharide-induced RAW 264.7 macrophages. *Biochem Biophys Res Commun* 236:655-660.

525. Goldring, C. E., R. Narayanan, P. Lagadec, and J. F. Jeannin. 1995. Transcriptional inhibition of the inducible nitric oxide synthase gene by competitive binding of NF-kappa B/Rel proteins. *Biochem Biophys Res Commun* 209:73-79.

526. Ohmori, Y., R. D. Schreiber, and T. A. Hamilton. 1997. Synergy between interferon-gamma and tumor necrosis factor-alpha in transcriptional activation is mediated by cooperation between signal transducer and activator of transcription 1 and nuclear factor kappaB. *J Biol Chem* 272:14899-14907.

527. Paludan, S. R. 2000. Synergistic action of pro-inflammatory agents: cellular and molecular aspects. *J Leukoc Biol* 67:18-25.

528. Paludan, S. R., L. Malmgaard, S. Ellermann-Eriksen, L. Bosca, and S. C. Mogensen. 2001. Interferon (IFN)-gamma and Herpes simplex virus/tumor necrosis factor-alpha synergistically induce nitric oxide synthase 2 in macrophages through cooperative action of nuclear factor-kappa B and IFN regulatory factor-1. *Eur Cytokine Netw* 12:297-308.

529. Saura, M., C. Zaragoza, C. Bao, A. McMillan, and C. J. Lowenstein. 1999. Interaction of interferon regulatory factor-1 and nuclear factor kappaB during activation of inducible nitric oxide synthase transcription. *J Mol Biol* 289:459-471.

530. Kamijo, R., H. Harada, T. Matsuyama, M. Bosland, J. Gerecitano, D. Shapiro, J. Le, S. I. Koh, T. Kimura, S. J. Green, and et al. 1994. Requirement for transcription factor IRF-1 in NO synthase induction in macrophages. *Science* 263:1612-1615.
531. Martin, E., C. Nathan, and Q. W. Xie. 1994. Role of interferon regulatory factor 1 in induction of nitric oxide synthase. *J Exp Med* 180:977-984.
532. Gao, J., D. C. Morrison, T. J. Parmely, S. W. Russell, and W. J. Murphy. 1997. An interferon-gamma-activated site (GAS) is necessary for full expression of the mouse iNOS gene in response to interferon-gamma and lipopolysaccharide. *J Biol Chem* 272:1226-1230.
533. Darnell, J. E., Jr., I. M. Kerr, and G. R. Stark. 1994. Jak-STAT pathways and transcriptional activation in response to IFNs and other extracellular signaling proteins. *Science* 264:1415-1421.
534. Rein, T., M. Muller, and H. Zorbas. 1994. In vivo footprinting of the IRF-1 promoter: inducible occupation of a GAS element next to a persistent structural alteration of the DNA. *Nucleic Acids Res* 22:3033-3037.
535. Meraz, M. A., J. M. White, K. C. Sheehan, E. A. Bach, S. J. Rodig, A. S. Dighe, D. H. Kaplan, J. K. Riley, A. C. Greenlund, D. Campbell, K. Carver-Moore, R. N. DuBois, R. Clark, M. Aguet, and R. D. Schreiber. 1996. Targeted disruption of the Stat1 gene in mice reveals unexpected physiologic specificity in the JAK-STAT signaling pathway. *Cell* 84:431-442.
536. Lorsbach, R. B., W. J. Murphy, C. J. Lowenstein, S. H. Snyder, and S. W. Russell. 1993. Expression of the nitric oxide synthase gene in mouse macrophages activated for tumor cell killing. Molecular basis for the synergy between interferon-gamma and lipopolysaccharide. *J Biol Chem* 268:1908-1913.
537. Drapier, J. C., J. Wietzerbin, and J. B. Hibbs, Jr. 1988. Interferon-gamma and tumor necrosis factor induce the L-arginine-dependent cytotoxic effector mechanism in murine macrophages. *Eur J Immunol* 18:1587-1592.
538. Sekine, N., T. Ishikawa, T. Okazaki, M. Hayashi, C. B. Wollheim, and T. Fujita. 2000. Synergistic activation of NF-kappab and inducible isoform of nitric oxide synthase induction by interferon-gamma and tumor necrosis factor-alpha in INS-1 cells. *J Cell Physiol* 184:46-57.
539. Teng, X., H. Zhang, C. Snead, and J. D. Catravas. 2002. Molecular mechanisms of iNOS induction by IL-1 beta and IFN-gamma in rat aortic smooth muscle cells. *Am J Physiol Cell Physiol* 282:C144-152.
540. Nomura, Y. 2001. NF-kappaB activation and IkappaB alpha dynamism involved in iNOS and chemokine induction in astroglial cells. *Life Sci* 68:1695-1701.
541. Jaramillo, M., P. H. Naccache, and M. Olivier. 2004. Monosodium urate crystals synergize with IFN-gamma to generate macrophage nitric oxide: involvement of extracellular signal-regulated kinase 1/2 and NF-kappa B. *J Immunol* 172:5734-5742.
542. Vodovotz, Y., and C. Bogdan. 1994. Control of nitric oxide synthase expression by transforming growth factor-beta: implications for homeostasis. *Prog Growth Factor Res* 5:341-351.
543. Galea, E., D. J. Reis, and D. L. Feinstein. 1994. Cloning and expression of inducible nitric oxide synthase from rat astrocytes. *J Neurosci Res* 37:406-414.
544. Park, S. K., and S. Murphy. 1996. Nitric oxide synthase type II mRNA stability is translation- and transcription-dependent. *J Neurochem* 67:1766-1769.
545. Felley-Bosco, E., F. C. Bender, F. Courjault-Gautier, C. Bron, and A. F. Quest. 2000. Caveolin-1 down-regulates inducible nitric oxide synthase via the proteasome pathway in human colon carcinoma cells. *Proc Natl Acad Sci U S A* 97:14334-14339.
546. Musial, A., and N. T. Eissa. 2001. Inducible nitric-oxide synthase is regulated by the proteasome degradation pathway. *J Biol Chem* 276:24268-24273.
547. Tochio, H., S. Ohki, Q. Zhang, M. Li, and M. Zhang. 1998. Solution structure of a protein inhibitor of neuronal nitric oxide synthase. *Nat Struct Biol* 5:965-969.
548. Ratovitski, E. A., C. Bao, R. A. Quick, A. McMillan, C. Kozlovsky, and C. J. Lowenstein. 1999. An inducible nitric-oxide synthase (NOS)-associated protein inhibits NOS dimerization and activity. *J Biol Chem* 274:30250-30257.
549. Werner-Felmayer, G., G. Golderer, and E. R. Werner. 2002. Tetrahydrobiopterin biosynthesis, utilization and pharmacological effects. *Curr Drug Metab* 3:159-173.
550. Michel, T., and O. Feron. 1997. Nitric oxide synthases: which, where, how, and why? *J Clin Invest* 100:2146-2152.
551. Morales-Ruiz, M., M. J. Lee, S. Zollner, J. P. Gratton, R. Scotland, I. Shiojima, K. Walsh, T. Hla, and W. C. Sessa. 2001. Sphingosine 1-phosphate activates Akt, nitric oxide production, and chemotaxis through a Gi protein/phosphoinositide 3-kinase pathway in endothelial cells. *J Biol Chem* 276:19672-19677.
552. Pan, J., K. L. Burgher, A. M. Szczepanik, and G. E. Ringheim. 1996. Tyrosine phosphorylation of inducible nitric oxide synthase: implications for potential post-translational regulation. *Biochem J* 314 (Pt 3):889-894.
553. Salh, B., R. Wagey, A. Marotta, J. S. Tao, and S. Pelech. 1998. Activation of phosphatidylinositol 3-kinase, protein kinase B, and p70 S6 kinases in lipopolysaccharide-stimulated Raw 264.7 cells: differential effects of rapamycin, Ly294002, and wortmannin on nitric oxide production. *J Immunol* 161:6947-6954.
554. Hattori, Y., E. B. Campbell, and S. S. Gross. 1994. Argininosuccinate synthetase mRNA and activity are induced by immunostimulants in vascular smooth muscle. Role in the regeneration or arginine for nitric oxide synthesis. *J Biol Chem* 269:9405-9408.

555. Nussler, A. K., T. R. Billiar, Z. Z. Liu, and S. M. Morris, Jr. 1994. Coinduction of nitric oxide synthase and argininosuccinate synthetase in a murine macrophage cell line. Implications for regulation of nitric oxide production. *J Biol Chem* 269:1257-1261.
556. Closs, E. I., J. S. Scheld, M. Sharafi, and U. Forstermann. 2000. Substrate supply for nitric-oxide synthase in macrophages and endothelial cells: role of cationic amino acid transporters. *Mol Pharmacol* 57:68-74.
557. Wu, G., and S. M. Morris, Jr. 1998. Arginine metabolism: nitric oxide and beyond. *Biochem J* 336 (Pt 1):1-17.
558. Munder, M., K. Eichmann, J. M. Moran, F. Centeno, G. Soler, and M. Modolell. 1999. Th1/Th2-regulated expression of arginase isoforms in murine macrophages and dendritic cells. *J Immunol* 163:3771-3777.
559. Gotoh, T., and M. Mori. 1999. Arginase II downregulates nitric oxide (NO) production and prevents NO-mediated apoptosis in murine macrophage-derived RAW 264.7 cells. *J Cell Biol* 144:427-434.
560. Mackay, C. R. 2001. Chemokines: immunology's high impact factors. *Nat Immunol* 2:95-101.
561. Olson, T. S., and K. Ley. 2002. Chemokines and chemokine receptors in leukocyte trafficking. *Am J Physiol Regul Integr Comp Physiol* 283:R7 28.
562. Clark-Lewis, I., K. S. Kim, K. Rajarathnam, J. H. Gong, B. Dewald, B. Moser, M. Baggiolini, and B. D. Sykes. 1995. Structure-activity relationships of chemokines. *J Leukoc Biol* 57:703-711.
563. Murphy, P. M., M. Baggiolini, I. F. Charo, C. A. Hebert, R. Horuk, K. Matsushima, L. H. Miller, J. J. Oppenheim, and C. A. Power. 2000. International union of pharmacology. XXII. Nomenclature for chemokine receptors. *Pharmacol Rev* 52:145-176.
564. Zlotnik, A., and O. Yoshie. 2000. Chemokines: a new classification system and their role in immunity. *Immunity* 12:121-127.
565. Murphy, P. M. 1994. The molecular biology of leukocyte chemoattractant receptors. *Annu Rev Immunol* 12:593-633.
566. Murdoch, C., and A. Finn. 2000. Chemokine receptors and their role in inflammation and infectious diseases. *Blood* 95:3032-3043.
567. McColl, S. R. 2002. Chemokines and dendritic cells: a crucial alliance. *Immunol Cell Biol* 80:489-496.
568. Holmes, W. E., J. Lee, W. J. Kuang, G. C. Rice, and W. I. Wood. 1991. Structure and functional expression of a human interleukin-8 receptor. *Science* 253:1278-1280.
569. Loetscher, M., T. Geiser, T. O'Reilly, R. Zwahlen, M. Baggiolini, and B. Moser. 1994. Cloning of a human seven-transmembrane domain receptor, LESTR, that is highly expressed in leukocytes. *J Biol Chem* 269:232-237.
570. Aragay, A. M., M. Mellado, J. M. Frade, A. M. Martin, M. C. Jimenez-Sainz, A. C. Martinez, and F. Mayor, Jr. 1998. Monocyte chemoattractant protein-1-induced CCR2B receptor desensitization mediated by the G protein-coupled receptor kinase 2. *Proc Natl Acad Sci U S A* 95:2985-2990.
571. Stanton, K. J., M. B. Frewin, and P. W. Gudewicz. 1999. Heterologous desensitization of IL-8-mediated chemotaxis in human neutrophils by a cell-binding fragment of fibronectin. *J Leukoc Biol* 65:515-522.
572. Sallusto, F., D. Lenig, C. R. Mackay, and A. Lanzavecchia. 1998. Flexible programs of chemokine receptor expression on human polarized T helper 1 and 2 lymphocytes. *J Exp Med* 187:875-883.
573. Jinquan, T., S. Quan, G. Feili, C. G. Larsen, and K. Thestrup-Pedersen. 1999. Eotaxin activates T cells to chemotaxis and adhesion only if induced to express CCR3 by IL-2 together with IL-4. *J Immunol* 162:4285-4292.
574. Loetscher, P., M. Seitz, M. Baggiolini, and B. Moser. 1996. Interleukin-2 regulates CC chemokine receptor expression and chemotactic responsiveness in T lymphocytes. *J Exp Med* 184:569-577.
575. Syrbe, U., J. Siveke, and A. Hamann. 1999. Th1/Th2 subsets: distinct differences in homing and chemokine receptor expression? *Springer Semin Immunopathol* 21:263-285.
576. Matte, C., and M. Olivier. 2002. Leishmania-induced cellular recruitment during the early inflammatory response: modulation of proinflammatory mediators. *J Infect Dis* 185:673-681.
577. Huang, L., I. Bosch, W. Hofmann, J. Sodroski, and A. B. Pardee. 1998. Tat protein induces human immunodeficiency virus type 1 (HIV-1) coreceptors and promotes infection with both macrophage-tropic and T-lymphotropic HIV-1 strains. *J Virol* 72:8952-8960.
578. Fonteneau, J. F., M. Gilliet, M. Larsson, I. Dasilva, C. Munz, Y. J. Liu, and N. Bhardwaj. 2003. Activation of influenza virus-specific CD4+ and CD8+ T cells: a new role for plasmacytoid dendritic cells in adaptive immunity. *Blood* 101:3520-3526.
579. Shaw, J. H., V. R. Grund, L. Durling, and H. D. Caldwell. 2001. Expression of genes encoding Th1 cell-activating cytokines and lymphoid homing chemokines by chlamydia-pulsed dendritic cells correlates with protective immunizing efficacy. *Infect Immun* 69:4667-4672.
580. Sallusto, F., E. Kremmer, B. Palermo, A. Hoy, P. Ponath, S. Qin, R. Forster, M. Lipp, and A. Lanzavecchia. 1999. Switch in chemokine receptor expression upon TCR stimulation reveals novel homing potential for recently activated T cells. *Eur J Immunol* 29:2037-2045.
581. Sallusto, F., B. Palermo, D. Lenig, M. Miettinen, S. Matikainen, I. Julkunen, R. Forster, R. Burgstahler, M. Lipp, and A. Lanzavecchia. 1999. Distinct patterns and kinetics of chemokine production regulate dendritic cell function. *Eur J Immunol* 29:1617-1625.
582. Thelen, M. 2001. Dancing to the tune of chemokines. *Nat Immunol* 2:129-134.

583. Kuang, Y., Y. Wu, H. Jiang, and D. Wu. 1996. Selective G protein coupling by C-C chemokine receptors. *J Biol Chem* 271:3975-3978.

584. Wu, D., G. J. LaRosa, and M. I. Simon. 1993. G protein-coupled signal transduction pathways for interleukin-8. *Science* 261:101-103.

585. Stephens, L., A. Smrcka, F. T. Cooke, T. R. Jackson, P. C. Sternweis, and P. T. Hawkins. 1994. A novel phosphoinositide 3 kinase activity in myeloid-derived cells is activated by G protein beta gamma subunits. *Cell* 77:83-93.

586. Meili, R., C. Ellsworth, S. Lee, T. B. Reddy, H. Ma, and R. A. Firtel. 1999. Chemoattractant-mediated transient activation and membrane localization of Akt/PKB is required for efficient chemotaxis to cAMP in Dictyostelium. *EMBO J* 18:2092-2105.

587. Sasaki, T., J. Irie-Sasaki, R. G. Jones, A. J. Oliveira-dos-Santos, W. L. Stanford, B. Bolon, A. Wakeham, A. Itie, D. Bouchard, I. Kozieradzki, N. Joza, T. W. Mak, P. S. Ohashi, A. Suzuki, and J. M. Penninger. 2000. Function of PI3Kgamma in thymocyte development, T cell activation, and neutrophil migration. *Science* 287:1040-1046.

588. Bacon, K. B., M. C. Szabo, H. Yssel, J. B. Bolen, and T. J. Schall. 1996. RANTES induces tyrosine kinase activity of stably complexed p125FAK and ZAP-70 in human T cells. *J Exp Med* 184:873-882.

589. Turner, S. J., J. Domin, M. D. Waterfield, S. G. Ward, and J. Westwick. 1998. The CC chemokine monocyte chemotactic peptide-1 activates both the class I p85/p110 phosphatidylinositol 3-kinase and the class II PI3K-C2alpha. *J Biol Chem* 273:25987-25995.

590. Ma, Y. C., J. Huang, S. Ali, W. Lowry, and X. Y. Huang. 2000. Src tyrosine kinase is a novel direct effector of G proteins. *Cell* 102:635-646.

591. Kampen, G. T., S. Stafford, L. K. Poulsen, and R. Alam. 1998. The MAPK ERK2 is activated through CCR3 and is essential for eotaxin-stimulated eosinophil chemotaxis. *J. Allergy Exp. Immunol.* 101:223.

592. Rossi, D., and A. Zlotnik. 2000. The biology of chemokines and their receptors. *Annu Rev Immunol* 18:217-242.

593. Springer, T. A. 1994. Traffic signals for lymphocyte recirculation and leukocyte emigration: the multistep paradigm. *Cell* 76:301-314.

594. Sallusto, F., C. R. Mackay, and A. Lanzavecchia. 2000. The role of chemokine receptors in primary, effector, and memory immune responses. *Annu Rev Immunol* 18:593-620.

595. Constantin, G., M. Majeed, C. Giagulli, L. Piccio, J. Y. Kim, E. C. Butcher, and C. Laudanna. 2000. Chemokines trigger immediate beta2 integrin affinity and mobility changes: differential regulation and roles in lymphocyte arrest under flow. *Immunity* 13:759-769.

596. Borish, L. C., and J. W. Steinke. 2003. 2. Cytokines and chemokines. *J Allergy Clin Immunol* 111:S460-475.

597. Foxman, E. F., J. J. Campbell, and E. C. Butcher. 1997. Multistep navigation and the combinatorial control of leukocyte chemotaxis. *J Cell Biol* 139:1349-1360.

598. Luster, A. D. 2002. The role of chemokines in linking innate and adaptive immunity. *Curr Opin Immunol* 14:129-135.

599. Gallucci, S., and P. Matzinger. 2001. Danger signals: SOS to the immune system. *Curr Opin Immunol* 13:114-119.

600. Racoosin, E. L., and S. M. Beverley. 1997. Leishmania major: promastigotes induce expression of a subset of chemokine genes in murine macrophages. *Exp Parasitol* 85:283-295.

601. Re, F., and J. L. Strominger. 2001. Toll-like receptor 2 (TLR2) and TLR4 differentially activate human dendritic cells. *J Biol Chem* 276:37692-37699.

602. Dieu, M. C., B. Vanbervliet, A. Vicari, J. M. Bridon, E. Oldham, S. Ait-Yahia, F. Briere, A. Zlotnik, S. Lebecque, and C. Caux. 1998. Selective recruitment of immature and mature dendritic cells by distinct chemokines expressed in different anatomic sites. *J Exp Med* 188:373-386.

603. Sozzani, S., P. Allavena, G. D'Amico, W. Luini, G. Bianchi, M. Kataura, T. Imai, O. Yoshie, R. Bonecchi, and A. Mantovani. 1998. Differential regulation of chemokine receptors during dendritic cell maturation: a model for their trafficking properties. *J Immunol* 161:1083-1086.

604. Sozzani, S., P. Allavena, A. Vecchi, and A. Mantovani. 2000. Chemokines and dendritic cell traffic. *J Clin Immunol* 20:151-160.

605. Michelsen, K. S., A. Aicher, M. Mohaupt, T. Hartung, S. Dimmeler, C. J. Kirschning, and R. R. Schumann. 2001. The role of toll-like receptors (TLRs) in bacteria-induced maturation of murine dendritic cells (DCS). Peptidoglycan and lipoteichoic acid are inducers of DC maturation and require TLR2. *J Biol Chem* 276:25680-25686.

606. Luther, S. A., H. L. Tang, P. L. Hyman, A. G. Farr, and J. G. Cyster. 2000. Coexpression of the chemokines ELC and SLC by T zone stromal cells and deletion of the ELC gene in the plt/plt mouse. *Proc Natl Acad Sci U S A* 97:12694-12699.

607. Schaerli, P., K. Willimann, A. B. Lang, M. Lipp, P. Loetscher, and B. Moser. 2000. CXC chemokine receptor 5 expression defines follicular homing T cells with B cell helper function. *J Exp Med* 192:1553-1562.

608. Salazar-Mather, T. P., J. S. Orange, and C. A. Biron. 1998. Early murine cytomegalovirus (MCMV) infection induces liver natural killer (NK) cell inflammation and protection through macrophage inflammatory protein 1alpha (MIP-1alpha)-dependent pathways. *J Exp Med* 187:1-14.
609. Aliberti, J., C. Reis e Sousa, M. Schito, S. Hieny, T. Wells, G. B. Huffnagle, and A. Sher. 2000. CCR5 provides a signal for microbial induced production of IL-12 by CD8 alpha+ dendritic cells. *Nat Immunol* 1:83-87.
610. Khan, I. A., J. A. MacLean, F. S. Lee, L. Casciotti, E. DeHaan, J. D. Schwartzman, and A. D. Luster. 2000. IP-10 is critical for effector T cell trafficking and host survival in Toxoplasma gondii infection. *Immunity* 12:483-494.
611. Wagner, L., O. O. Yang, E. A. Garcia-Zepeda, Y. Ge, S. A. Kalams, B. D. Walker, M. S. Pasternack, and A. D. Luster. 1998. Beta-chemokines are released from HIV-1-specific cytolytic T-cell granules complexed to proteoglycans. *Nature* 391:908-911.
612. Gerard, C., and B. J. Rollins. 2001. Chemokines and disease. *Nat Immunol* 2:108-115.
613. Christopherson, K., 2nd, and R. Hromas. 2001. Chemokine regulation of normal and pathologic immune responses. *Stem Cells* 19:388-396.
614. Katoh, S., N. Matsumoto, K. Fukushima, H. Mukae, J. I. Kadota, S. Kohno, and S. Matsukura. 2000. Elevated chemokine levels in bronchoalveolar lavage fluid of patients with eosinophilic pneumonia. *J Allergy Clin Immunol* 106:730-736.
615. Brenier-Pinchart, M. P., H. Pelloux, D. Derouich-Guergour, and P. Ambroise-Thomas. 2001. Chemokines in host-protozoan-parasite interactions. *Trends Parasitol* 17:292-296.
616. Shaio, M. F., P. R. Lin, J. Y. Liu, and K. D. Yang. 1995. Generation of interleukin-8 from human monocytes in response to Trichomonas vaginalis stimulation. *Infect Immun* 63:3864-3870.
617. Seydel, K. B., E. Li, P. E. Swanson, and S. L. Stanley, Jr. 1997. Human intestinal epithelial cells produce proinflammatory cytokines in response to infection in a SCID mouse-human intestinal xenograft model of amebiasis. *Infect Immun* 65:1631-1639.
618. Aliberti, J. C., F. S. Machado, J. T. Souto, A. P. Campanelli, M. M. Teixeira, R. T. Gazzinelli, and J. S. Silva. 1999. beta-Chemokines enhance parasite uptake and promote nitric oxide-dependent microbiostatic activity in murine inflammatory macrophages infected with Trypanosoma cruzi. *Infect Immun* 67:4819-4826.
619. Villalta, F., Y. Zhang, K. E. Bibb, J. C. Kappes, and M. F. Lima. 1998. The cysteine-cysteine family of chemokines RANTES, MIP-1alpha, and MIP-1beta induce trypanocidal activity in human macrophages via nitric oxide. *Infect Immun* 66:4690-4695.
620. Sharafeldin, A., R. Eltayeb, M. Pashenkov, and M. Bakhiet. 2000. Chemokines are produced in the brain early during the course of experimental African trypanosomiasis. *J Neuroimmunol* 103:165-170.
621. Amichay, D., R. T. Gazzinelli, G. Karupiah, T. R. Moench, A. Sher, and J. M. Farber. 1996. Genes for chemokines MuMig and Crg-2 are induced in protozoan and viral infections in response to IFN-gamma with patterns of tissue expression that suggest nonredundant roles in vivo. *J Immunol* 157:4511-4520.
622. Denney, C. F., L. Eckmann, and S. L. Reed. 1999. Chemokine secretion of human cells in response to Toxoplasma gondii infection. *Infect Immun* 67:1547-1552.
623. Bliss, S. K., A. J. Marshall, Y. Zhang, and E. Y. Denkers. 1999. Human polymorphonuclear leukocytes produce IL-12, TNF-alpha, and the chemokines macrophage-inflammatory protein-1 alpha and -1 beta in response to Toxoplasma gondii antigens. *J Immunol* 162:7369-7375.
624. Mannheimer, S. B., J. Hariprashad, M. Y. Stoeckle, and H. W. Murray. 1996. Induction of macrophage antiprotozoal activity by monocyte chemotactic and activating factor. *FEMS Immunol Med Microbiol* 14:59-61.
625. Ritter, U., H. Moll, T. Laskay, E. Brocker, O. Velazco, I. Becker, and R. Gillitzer. 1996. Differential expression of chemokines in patients with localized and diffuse cutaneous American leishmaniasis. *J Infect Dis* 173:699-709.
626. Vester, B., K. Muller, W. Solbach, and T. Laskay. 1999. Early gene expression of NK cell-activating chemokines in mice resistant to Leishmania major. *Infect Immun* 67:3155-3159.
627. Gu, L., S. Tseng, R. M. Horner, C. Tam, M. Loda, and B. J. Rollins. 2000. Control of TH2 polarization by the chemokine monocyte chemoattractant protein-1. *Nature* 404:407-411.
628. Sato, N., S. K. Ahuja, M. Quinones, V. Kostecki, R. L. Reddick, P. C. Melby, W. A. Kuziel, and S. S. Ahuja. 2000. CC chemokine receptor (CCR)2 is required for langerhans cell migration and localization of T helper cell type 1 (Th1)-inducing dendritic cells. Absence of CCR2 shifts the Leishmania major-resistant phenotype to a susceptible state dominated by Th2 cytokines, b cell outgrowth, and sustained neutrophilic inflammation. *J Exp Med* 192:205-218.
629. Cotterell, S. E., C. R. Engwerda, and P. M. Kaye. 1999. Leishmania donovani infection initiates T cell-independent chemokine responses, which are subsequently amplified in a T cell-dependent manner. *Eur J Immunol* 29:203-214.
630. Herwaldt, B. L. 1999. Leishmaniasis. *Lancet* 354:1191-1199.

631. Chaudhuri, A., J. Polyakova, V. Zbrzezna, K. Williams, S. Gulati, and A. O. Pogo. 1993. Cloning of glycoprotein D cDNA, which encodes the major subunit of the Duffy blood group system and the receptor for the Plasmodium vivax malaria parasite. *Proc Natl Acad Sci U S A* 90:10793-10797.

632. Hadley, T. J., and S. C. Peiper. 1997. From malaria to chemokine receptor: the emerging physiologic role of the Duffy blood group antigen. *Blood* 89:3077-3091.

633. Horuk, R., C. E. Chitnis, W. C. Darbonne, T. J. Colby, A. Rybicki, T. J. Hadley, and L. H. Miller. 1993. A receptor for the malarial parasite Plasmodium vivax: the erythrocyte chemokine receptor. *Science* 261:1182-1184.

634. Horuk, R. 1994. The interleukin-8-receptor family: from chemokines to malaria. *Immunol Today* 15:169-174.

635. Proudfoot, A. E. 2002. Chemokine receptors: multifaceted therapeutic targets. *Nat Rev Immunol* 2:106-115.

636. Onuffer, J. J., and R. Horuk. 2002. Chemokines, chemokine receptors and small-molecule antagonists: recent developments. *Trends Pharmacol Sci* 23:459-467.

637. Liang, M., C. Mallari, M. Rosser, H. P. Ng, K. May, S. Monahan, J. G. Bauman, I. Islam, A. Ghannam, B. Buckman, K. Shaw, G. P. Wei, W. Xu, Z. Zhao, E. Ho, J. Shen, H. Oanh, B. Subramanyam, R. Vergona, D. Taub, L. Dunning, S. Harvey, R. M. Snider, J. Hesselgesser, M. M. Morrissey, and H. D. Perez. 2000. Identification and characterization of a potent, selective, and orally active antagonist of the CC chemokine receptor-1. *J Biol Chem* 275:19000-19008.

638. Bruhl, H., J. Cihak, M. Stangassinger, D. Schlondorff, and M. Mack. 2001. Depletion of CCR5-expressing cells with bispecific antibodies and chemokine toxins: a new strategy in the treatment of chronic inflammatory diseases and HIV. *J Immunol* 166:2420-2426.

639. Hancock, W. W., B. Lu, W. Gao, V. Csizmadia, K. Faia, J. A. King, S. T. Smiley, M. Ling, N. P. Gerard, and C. Gerard. 2000. Requirement of the chemokine receptor CXCR3 for acute allograft rejection. *J Exp Med* 192:1515-1520.

640. Youssef, S., G. Maor, G. Wildbaum, N. Grabie, A. Gour-Lavie, and N. Karin. 2000. C-C chemokine-encoding DNA vaccines enhance breakdown of tolerance to their gene products and treat ongoing adjuvant arthritis. *J Clin Invest* 106:361-371.

641. von Andrian, U. H., and C. R. Mackay. 2000. T-cell function and migration. Two sides of the same coin. *N Engl J Med* 343:1020-1034.

642. Bonecchi, R., S. Sozzani, J. T. Stine, W. Luini, G. D'Amico, P. Allavena, D. Chantry, and A. Mantovani. 1998. Divergent effects of interleukin-4 and interferon-gamma on macrophage-derived chemokine production: an amplification circuit of polarized T helper 2 responses. *Blood* 92:2668-2671.

643. O'Neill, L. A. 2002. Toll-like receptor signal transduction and the tailoring of innate immunity: a role for Mal? *Trends Immunol* 23:296-300.

644. Akira, S., K. Takeda, and T. Kaisho. 2001. Toll-like receptors: critical proteins linking innate and acquired immunity. *Nat Immunol* 2:675-680.

645. Hirschfeld, M., J. J. Weis, V. Toshchakov, C. A. Salkowski, M. J. Cody, D. C. Ward, N. Qureshi, S. M. Michalek, and S. N. Vogel. 2001. Signaling by toll-like receptor 2 and 4 agonists results in differential gene expression in murine macrophages. *Infect Immun* 69:1477-1482.

646. Akhtar, M., J. L. Watson, A. Nazli, and D. M. McKay. 2003. Bacterial DNA evokes epithelial IL-8 production by a MAPK-dependent, NF-kappaB-independent pathway. *FASEB J* 17:1319-1321.

647. Sabroe, I., L. R. Prince, E. C. Jones, M. J. Horsburgh, S. J. Foster, S. N. Vogel, S. K. Dower, and M. K. Whyte. 2003. Selective roles for Toll-like receptor (TLR)2 and TLR4 in the regulation of neutrophil activation and life span. *J Immunol* 170:5268-5275.

648. Tamai, R., S. Sugawara, O. Takeuchi, S. Akira, and H. Takada. 2003. Synergistic effects of lipopolysaccharide and interferon-gamma in inducing interleukin-8 production in human monocytic THP-1 cells is accompanied by up-regulation of CD14, Toll-like receptor 4, MD-2 and MyD88 expression. *J Endotoxin Res* 9:145-153.

649. Proost, P., A. K. Vynckier, F. Mahieu, W. Put, B. Grillet, S. Struyf, A. Wuyts, G. Opdenakker, and J. Van Damme. 2003. Microbial Toll-like receptor ligands differentially regulate CXCL10/IP-10 expression in fibroblasts and mononuclear leukocytes in synergy with IFN-gamma and provide a mechanism for enhanced synovial chemokine levels in septic arthritis. *Eur J Immunol* 33:3146-3153.

650. Smith, M. F., Jr., A. Mitchell, G. Li, S. Ding, A. M. Fitzmaurice, K. Ryan, S. Crowe, and J. B. Goldberg. 2003. Toll-like receptor (TLR) 2 and TLR5, but not TLR4, are required for Helicobacter pylori-induced NF-kappa B activation and chemokine expression by epithelial cells. *J Biol Chem* 278:32552-32560.

651. Fisette, P. L., S. Ram, J. M. Andersen, W. Guo, and R. R. Ingalls. 2003. The Lip lipoprotein from Neisseria gonorrhoeae stimulates cytokine release and NF-kappaB activation in epithelial cells in a Toll-like receptor 2-dependent manner. *J Biol Chem* 278:46252-46260.

652. Kusumoto, Y., H. Hirano, K. Saitoh, S. Yamada, M. Takedachi, T. Nozaki, Y. Ozawa, Y. Nakahira, T. Saho, H. Ogo, Y. Shimabukuro, H. Okada, and S. Murakami. 2004. Human gingival epithelial cells produce chemotactic factors interleukin-8 and monocyte chemoattractant protein-1 after stimulation with Porphyromonas gingivalis via toll-like receptor 2. *J Periodontol* 75:370-379.

653. Zhang, J., K. Xu, B. Ambati, and F. S. Yu. 2003. Toll-like receptor 5-mediated corneal epithelial inflammatory responses to Pseudomonas aeruginosa flagellin. *Invest Ophthalmol Vis Sci* 44:4247-4254.

654. Means, T. K., F. Hayashi, K. D. Smith, A. Aderem, and A. D. Luster. 2003. The Toll-like receptor 5 stimulus bacterial flagellin induces maturation and chemokine production in human dendritic cells. *J Immunol* 170:5165-5175.

655. Mempel, M., V. Voelcker, G. Kollisch, C. Plank, R. Rad, M. Gerhard, C. Schnopp, P. Fraunberger, A. K. Walli, J. Ring, D. Abeck, and M. Ollert. 2003. Toll-like receptor expression in human keratinocytes: nuclear factor kappaB controlled gene activation by Staphylococcus aureus is toll-like receptor 2 but not toll-like receptor 4 or platelet activating factor receptor dependent. *J Invest Dermatol* 121:1389-1396.

656. Meier, A., C. J. Kirschning, T. Nikolaus, H. Wagner, J. Heesemann, and F. Ebel. 2003. Toll-like receptor (TLR) 2 and TLR4 are essential for Aspergillus-induced activation of murine macrophages. *Cell Microbiol* 5:561-570.

657. Villamon, E., D. Gozalbo, P. Roig, J. E. O'Connor, D. Fradelizi, and M. L. Gil. 2004. Toll-like receptor-2 is essential in murine defenses against Candida albicans infections. *Microbes Infect* 6:1-7.

658. De Veer, M. J., J. M. Curtis, T. M. Baldwin, J. A. DiDonato, A. Sexton, M. J. McConville, E. Handman, and L. Schofield. 2003. MyD88 is essential for clearance of Leishmania major: possible role for lipophosphoglycan and Toll-like receptor 2 signaling. *Eur J Immunol* 33:2822-2831.

659. Ouaissi, A., E. Guilvard, Y. Delneste, G. Caron, G. Magistrelli, N. Herbault, N. Thieblemont, and P. Jeannin. 2002. The Trypanosoma cruzi Tc52-released protein induces human dendritic cell maturation, signals via Toll-like receptor 2, and confers protection against lethal infection. *J Immunol* 168:6366-6374.

660. Coelho, P. S., A. Klein, A. Talvani, S. F. Coutinho, O. Takeuchi, S. Akira, J. S. Silva, H. Canizzaro, R. T. Gazzinelli, and M. M. Teixeira. 2002. Glycosylphosphatidylinositol-anchored mucin-like glycoproteins isolated from Trypanosoma cruzi trypomastigotes induce in vivo leukocyte recruitment dependent on MCP-1 production by IFN-gamma-primed-macrophages. *J Leukoc Biol* 71:837-844.

661. Kurt-Jones, E. A., M. Chan, S. Zhou, J. Wang, G. Reed, R. Bronson, M. M. Arnold, D. M. Knipe, and R. W. Finberg. 2004. Herpes simplex virus 1 interaction with Toll-like receptor 2 contributes to lethal encephalitis. *Proc Natl Acad Sci U S A* 101:1315-1320.

662. Monick, M. M., T. O. Yarovinsky, L. S. Powers, N. S. Butler, A. B. Carter, G. Gudmundsson, and G. W. Hunninghake. 2003. Respiratory syncytial virus up-regulates TLR4 and sensitizes airway epithelial cells to endotoxin. *J Biol Chem* 278:53035-53044.

663. Smiley, S. T., J. A. King, and W. W. Hancock. 2001. Fibrinogen stimulates macrophage chemokine secretion through toll-like receptor 4. *J Immunol* 167:2887-2894.

664. Devaney, J. M., C. M. Greene, C. C. Taggart, T. P. Carroll, S. J. O'Neill, and N. G. McElvaney. 2003. Neutrophil elastase up-regulates interleukin-8 via toll-like receptor 4. *FEBS Lett* 544:129-132.

665. Roebuck, K. A., L. R. Carpenter, V. Lakshminarayanan, S. M. Page, J. N. Moy, and L. L. Thomas. 1999. Stimulus-specific regulation of chemokine expression involves differential activation of the redox-responsive transcription factors AP-1 and NF-kappaB. *J Leukoc Biol* 65:291-298.

666. Ping, D., P. L. Jones, and J. M. Boss. 1996. TNF regulates the in vivo occupancy of both distal and proximal regulatory regions of the MCP-1/JE gene. *Immunity* 4:455-469.

667. Grove, M., and M. Plumb. 1993. C/EBP, NF-kappa B, and c-Ets family members and transcriptional regulation of the cell-specific and inducible macrophage inflammatory protein 1 alpha immediate-early gene. *Mol Cell Biol* 13:5276-5289.

668. Widmer, U., K. R. Manogue, A. Cerami, and B. Sherry. 1993. Genomic cloning and promoter analysis of macrophage inflammatory protein (MIP)-2, MIP-1 alpha, and MIP-1 beta, members of the chemokine superfamily of proinflammatory cytokines. *J Immunol* 150:4996-5012.

669. Stein, B., and A. S. Baldwin, Jr. 1993. Distinct mechanisms for regulation of the interleukin-8 gene involve synergism and cooperativity between C/EBP and NF-kappa B. *Mol Cell Biol* 13:7191-7198.

670. Martin, T., P. M. Cardarelli, G. C. Parry, K. A. Felts, and R. R. Cobb. 1997. Cytokine induction of monocyte chemoattractant protein-1 gene expression in human endothelial cells depends on the cooperative action of NF-kappa B and AP-1. *Eur J Immunol* 27:1091-1097.

671. Wung, B. S., J. J. Cheng, H. J. Hsieh, Y. J. Shyy, and D. L. Wang. 1997. Cyclic strain-induced monocyte chemotactic protein-1 gene expression in endothelial cells involves reactive oxygen species activation of activator protein 1. *Circ Res* 81:1-7.

672. Lakshminarayanan, V., E. A. Drab-Weiss, and K. A. Roebuck. 1998. H_2O_2 and tumor necrosis factor-alpha induce differential binding of the redox-responsive transcription factors AP-1 and NF-kappaB to the interleukin-8 promoter in endothelial and epithelial cells. *J Biol Chem* 273:32670-32678.

673. Proffitt, J., G. Crabtree, M. Grove, P. Daubersies, B. Bailleul, E. Wright, and M. Plumb. 1995. An ATF/CREB-binding site is essential for cell-specific and inducible transcription of the murine MIP-1 beta cytokine gene. *Gene* 152:173-179.

674. Boehlk, S., S. Fessele, A. Mojaat, N. G. Miyamoto, T. Werner, E. L. Nelson, D. Schlondorff, and P. J. Nelson. 2000. ATF and Jun transcription factors, acting through an Ets/CRE promoter module, mediate lipopolysaccharide inducibility of the chemokine RANTES in monocytic Mono Mac 6 cells. *Eur J Immunol* 30:1102-1112.

675. Grewal, T., M. Boudreau, M. Roy, A. Chamberland, C. Lefebvre, J. Lavigne, J. Davignon, and A. Minnich. 1998. Expression of gamma-IFN responsive genes in scavenger receptor over-expressing monocytes is associated with xanthomatosis. *Atherosclerosis* 138:335-345.
676. Nazar, A. S., G. Cheng, H. S. Shin, P. N. Brothers, S. Dhib-Jalbut, M. L. Shin, and P. Vanguri. 1997. Induction of IP-10 chemokine promoter by measles virus: comparison with interferon-gamma shows the use of the same response element but with differential DNA-protein binding profiles. *J Neuroimmunol* 77:116-127.
677. Casola, A., R. P. Garofalo, H. Haeberle, T. F. Elliott, R. Lin, M. Jamaluddin, and A. R. Brasier. 2001. Multiple cis regulatory elements control RANTES promoter activity in alveolar epithelial cells infected with respiratory syncytial virus. *J Virol* 75:6428-6439.
678. Richmond, A. 2002. Nf-kappa B, chemokine gene transcription and tumour growth. *Nat Rev Immunol* 2:664-674.
679. Zhou, Z. H., P. Chaturvedi, Y. L. Han, S. Aras, Y. S. Li, P. E. Kolattukudy, D. Ping, J. M. Boss, and R. M. Ransohoff. 1998. IFN-gamma induction of the human monocyte chemoattractant protein (hMCP)-1 gene in astrocytoma cells: functional interaction between an IFN-gamma-activated site and a GC-rich element. *J Immunol* 160:3908-3916.
680. Horton, M. R., S. Boodoo, and J. D. Powell. 2002. NF-kappa B activation mediates the cross-talk between extracellular matrix and interferon-gamma (IFN-gamma) leading to enhanced monokine induced by IFN-gamma (MIG) expression in macrophages. *J Biol Chem* 277:43757-43762.
681. Funakoshi, M., Y. Sonoda, K. Tago, S. Tominaga, and T. Kasahara. 2001. Differential involvement of p38 mitogen-activated protein kinase and phosphatidyl inositol 3-kinase in the IL-1-mediated NF-kappa B and AP-1 activation. *Int Immunopharmacol* 1:595-604.
682. Parhar, K., A. Ray, U. Steinbrecher, C. Nelson, and B. Salh. 2003. The p38 mitogen-activated protein kinase regulates interleukin-1beta-induced IL-8 expression via an effect on the IL-8 promoter in intestinal epithelial cells. *Immunology* 108:502-512.
683. Jaruga, B., F. Hong, R. Sun, S. Radaeva, and B. Gao. 2003. Crucial role of IL-4/STAT6 in T cell-mediated hepatitis: up-regulating eotaxins and IL-5 and recruiting leukocytes. *J Immunol* 171:3233-3244.
684. Zhou, L., A. Tan, S. Iasvovskaia, J. Li, A. Lin, and M. B. Hershenson. 2003. Ras and mitogen-activated protein kinase kinase kinase-1 coregulate activator protein-1- and nuclear factor-kappaB-mediated gene expression in airway epithelial cells. *Am J Respir Cell Mol Biol* 28:762-769.
685. Chen, X. L., Q. Zhang, R. Zhao, and R. M. Medford. 2004. Superoxide, H2O2, and iron are required for TNF-alpha-induced MCP-1 gene expression in endothelial cells: role of Rac1 and NADPH oxidase. *Am J Physiol Heart Circ Physiol* 286:H1001-1007.
686. De Keulenaer, G. W., M. Ushio-Fukai, Q. Yin, A. B. Chung, P. R. Lyons, N. Ishizaka, K. Rengarajan, W. R. Taylor, R. W. Alexander, and K. K. Griendling. 2000. Convergence of redox-sensitive and mitogen-activated protein kinase signaling pathways in tumor necrosis factor-alpha-mediated monocyte chemoattractant protein-1 induction in vascular smooth muscle cells. *Arterioscler Thromb Vasc Biol* 20:385-391.
687. Kim, D. S., J. H. Han, and H. J. Kwon. 2003. NF-kappaB and c-Jun-dependent regulation of macrophage inflammatory protein-2 gene expression in response to lipopolysaccharide in RAW 264.7 cells. *Mol Immunol* 40:633-643.
688. Wang, B., P. P. Cleary, H. Xu, and J. D. Li. 2003. Up-regulation of interleukin-8 by novel small cytoplasmic molecules of nontypeable Haemophilus influenzae via p38 and extracellular signal-regulated kinase pathways. *Infect Immun* 71:5523-5530.
689. Balasubramanian, A., R. K. Ganju, and J. E. Groopman. 2003. Hepatitis C virus and HIV envelope proteins collaboratively mediate interleukin-8 secretion through activation of p38 MAP kinase and SHP2 in hepatocytes. *J Biol Chem* 278:35755-35766.
690. Seo, J. H., J. W. Lim, H. Kim, and K. H. Kim. 2004. Helicobacter pylori in a Korean isolate activates mitogen-activated protein kinases, AP-1, and NF-kappaB and induces chemokine expression in gastric epithelial AGS cells. *Lab Invest* 84:49-62.
691. Kutsch, O., J. Oh, A. Nath, and E. N. Benveniste. 2000. Induction of the chemokines interleukin-8 and IP-10 by human immunodeficiency virus type 1 tat in astrocytes. *J Virol* 74:9214-9221.
692. Melchjorsen, J., and S. R. Paludan. 2003. Induction of RANTES/CCL5 by herpes simplex virus is regulated by nuclear factor kappa B and interferon regulatory factor 3. *J Gen Virol* 84:2491-2495.
693. Asai, Y., S. Hashimoto, K. Kujime, Y. Gon, K. Mizumura, K. Shimizu, and T. Horie. 2001. Amantadine inhibits RANTES production by influenzavirus-infected human bronchial epithelial cells. *Br J Pharmacol* 132:918-924.
694. Mori, N., N. Mukaida, D. W. Ballard, K. Matsushima, and N. Yamamoto. 1998. Human T-cell leukemia virus type I Tax transactivates human interleukin 8 gene through acting concurrently on AP-1 and nuclear factor-kappaB-like sites. *Cancer Res* 58:3993-4000.
695. Mendez-Samperio, P., J. Palma, and A. Vazquez. 2002. Signals involved in mycobacteria-induced CXCL-8 production by human monocytes. *J Interferon Cytokine Res* 22:189-197.
696. Song, C. H., J. S. Lee, S. H. Lee, K. Lim, H. J. Kim, J. K. Park, T. H. Paik, and E. K. Jo. 2003. Role of mitogen-activated protein kinase pathways in the production of tumor necrosis factor-alpha, interleukin-10,

and monocyte chemotactic protein-1 by Mycobacterium tuberculosis H37Rv-infected human monocytes. *J Clin Immunol* 23:194-201.

697. Kim, J. M., Y. K. Oh, Y. J. Kim, S. J. Cho, M. H. Ahn, and Y. J. Cho. 2001. Nuclear factor-kappa B plays a major role in the regulation of chemokine expression of HeLa cells in response to Toxoplasma gondii infection. *Parasitol Res* 87:758-763.

698. Brenier-Pinchart, M. P., H. Pelloux, J. Simon, J. Ricard, J. L. Bosson, and P. Ambroise-Thomas. 2000. Toxoplasma gondii induces the secretion of monocyte chemotactic protein-1 in human fibroblasts, in vitro. *Mol Cell Biochem* 209:79-87.

699. Ryu, J. S., J. H. Kang, S. Y. Jung, M. H. Shin, J. M. Kim, H. Park, and D. Y. Min. 2004. Production of interleukin-8 by human neutrophils stimulated with Trichomonas vaginalis. *Infect Immun* 72:1326-1332.

700. Omura, T., M. Yoshiyama, S. Kim, R. Matsumoto, Y. Nakamura, Y. Izumi, H. Ichijo, T. Sudo, K. Akioka, H. Iwao, K. Takeuchi, and J. Yoshikawa. 2004. Involvement of apoptosis signal-regulating kinase-1 on angiotensin II-induced monocyte chemoattractant protein-1 expression. *Arterioscler Thromb Vasc Biol* 24:270-275.

701. Kwon, H. J., and D. S. Kim. 2003. Regulation of macrophage inflammatory protein-2 gene expression in response to oligodeoxynucleotide containing CpG motifs in RAW 264.7 cells. *Biochem Biophys Res Commun* 308:608-613.

702. Gern, J. E., D. A. French, K. A. Grindle, R. A. Brockman-Schneider, S. Konno, and W. W. Busse. 2003. Double-stranded RNA induces the synthesis of specific chemokines by bronchial epithelial cells. *Am J Respir Cell Mol Biol* 28:731-737.

703. Liu, R., K. Aupperle, and R. Terkeltaub. 2001. Src family protein tyrosine kinase signaling mediates monosodium urate crystal-induced IL-8 expression by monocytic THP-1 cells. *J Leukoc Biol* 70:961-968.

704. Liu, R., M. O'Connell, K. Johnson, K. Pritzker, N. Mackman, and R. Terkeltaub. 2000. Extracellular signal-regulated kinase 1/extracellular signal-regulated kinase 2 mitogen-activated protein kinase signaling and activation of activator protein 1 and nuclear factor kappaB transcription factors play central roles in interleukin-8 expression stimulated by monosodium urate monohydrate and calcium pyrophosphate crystals in monocytic cells. *Arthritis Rheum* 43:1145-1155.

705. Kanellis, J., S. Watanabe, J. H. Li, D. H. Kang, P. Li, T. Nakagawa, A. Wamsley, D. Sheikh-Hamad, H. Y. Lan, L. Feng, and R. J. Johnson. 2003. Uric acid stimulates monocyte chemoattractant protein-1 production in vascular smooth muscle cells via mitogen-activated protein kinase and cyclooxygenase-2. *Hypertension* 41:1287-1293.

706. Yamada, M., S. Kim, K. Egashira, M. Takeya, T. Ikeda, O. Mimura, and H. Iwao. 2003. Molecular mechanism and role of endothelial monocyte chemoattractant protein-1 induction by vascular endothelial growth factor. *Arterioscler Thromb Vasc Biol* 23:1996-2001.

707. Denhardt, D. T. 1996. Signal-transducing protein phosphorylation cascades mediated by Ras/Rho proteins in the mammalian cell: the potential for multiplex signalling. *Biochem J* 318 (Pt 3):729-747.

708. Cobb, M. H. 1999. MAP kinase pathways. *Prog Biophys Mol Biol* 71:479-500.

709. Davis, R. J. 1994. MAPKs: new JNK expands the group. *Trends Biochem Sci* 19:470-473.

710. Paul, A., S. Wilson, C. M. Belham, C. J. Robinson, P. H. Scott, G. W. Gould, and R. Plevin. 1997. Stress-activated protein kinases: activation, regulation and function. *Cell Signal* 9:403-410.

711. Widmann, C., S. Gibson, M. B. Jarpe, and G. L. Johnson. 1999. Mitogen-activated protein kinase: conservation of a three-kinase module from yeast to human. *Physiol Rev* 79:143-180.

712. Frodin, M., N. Sekine, E. Roche, C. Filloux, M. Prentki, C. B. Wollheim, and E. Van Obberghen. 1995. Glucose, other secretagogues, and nerve growth factor stimulate mitogen-activated protein kinase in the insulin-secreting beta-cell line, INS-1. *J Biol Chem* 270:7882-7889.

713. Ludwig, S., K. Engel, A. Hoffmeyer, G. Sithanandam, B. Neufeld, D. Palm, M. Gaestel, and U. R. Rapp. 1996. 3pK, a novel mitogen-activated protein (MAP) kinase-activated protein kinase, is targeted by three MAP kinase pathways. *Mol Cell Biol* 16:6687-6697.

714. Minden, A., A. Lin, T. Smeal, B. Derijard, M. Cobb, R. Davis, and M. Karin. 1994. c-Jun N-terminal phosphorylation correlates with activation of the JNK subgroup but not the ERK subgroup of mitogen-activated protein kinases. *Mol Cell Biol* 14:6683-6688.

715. Minden, A., and M. Karin. 1997. Regulation and function of the JNK subgroup of MAP kinases. *Biochim Biophys Acta* 1333:F85-104.

716. Musti, A. M., M. Treier, and D. Bohmann. 1997. Reduced ubiquitin-dependent degradation of c-Jun after phosphorylation by MAP kinases. *Science* 275:400-402.

717. Karin, M. 1998. Mitogen-activated protein kinase cascades as regulators of stress responses. *Ann N Y Acad Sci* 851:139-146.

718. Whitmarsh, A. J., and R. J. Davis. 1996. Transcription factor AP-1 regulation by mitogen-activated protein kinase signal transduction pathways. *J Mol Med* 74:589-607.

719. Frost, J. A., H. Steen, P. Shapiro, T. Lewis, N. Ahn, P. E. Shaw, and M. H. Cobb. 1997. Cross-cascade activation of ERKs and ternary complex factors by Rho family proteins. *EMBO J* 16:6426-6438.

720. Pulverer, B. J., J. M. Kyriakis, J. Avruch, E. Nikolakaki, and J. R. Woodgett. 1991. Phosphorylation of c-jun mediated by MAP kinases. *Nature* 353:670-674.

721. Sgambato, V., P. Vanhoutte, C. Pages, M. Rogard, R. Hipskind, M. J. Besson, and J. Caboche. 1998. In vivo expression and regulation of Elk-1, a target of the extracellular-regulated kinase signaling pathway, in the adult rat brain. *J Neurosci* 18:214-226.

722. Alvarez, E., I. C. Northwood, F. A. Gonzalez, D. A. Latour, A. Seth, C. Abate, T. Curran, and R. J. Davis. 1991. Pro-Leu-Ser/Thr-Pro is a consensus primary sequence for substrate protein phosphorylation. Characterization of the phosphorylation of c-myc and c-jun proteins by an epidermal growth factor receptor threonine 669 protein kinase. *J Biol Chem* 266:15277-15285.

723. Xing, J., D. D. Ginty, and M. E. Greenberg. 1996. Coupling of the RAS-MAPK pathway to gene activation by RSK2, a growth factor-regulated CREB kinase. *Science* 273:959-963.

724. Baumann, B., C. K. Weber, J. Troppmair, S. Whiteside, A. Israel, U. R. Rapp, and T. Wirth. 2000. Raf induces NF-kappaB by membrane shuttle kinase MEKK1, a signaling pathway critical for transformation. *Proc Natl Acad Sci U S A* 97:4615-4620.

725. Nemoto, S., J. A. DiDonato, and A. Lin. 1998. Coordinate regulation of IkappaB kinases by mitogen-activated protein kinase kinase kinase 1 and NF-kappaB-inducing kinase. *Mol Cell Biol* 18:7336-7343.

726. Norris, J. L., and A. S. Baldwin, Jr. 1999. Oncogenic Ras enhances NF-kappaB transcriptional activity through Raf-dependent and Raf-independent mitogen-activated protein kinase signaling pathways. *J Biol Chem* 274:13841-13846.

727. Pahl, H. L. 1999. Activators and target genes of Rel/NF-kappaB transcription factors. *Oncogene* 18:6853-6866.

728. Govind, S. 1999. Control of development and immunity by rel transcription factors in Drosophila. *Oncogene* 18:6875-6887.

729. Chen, F. E., and G. Ghosh. 1999. Regulation of DNA binding by Rel/NF-kappaB transcription factors: structural views. *Oncogene* 18:6845-6852.

730. Schmitz, M. L., and P. A. Baeuerle. 1991. The p65 subunit is responsible for the strong transcription activating potential of NF-kappa B. *EMBO J* 10:3805-3817.

731. Karin, M., and Y. Ben-Neriah. 2000. Phosphorylation meets ubiquitination: the control of NF-[kappa]B activity. *Annu Rev Immunol* 18:621-663.

732. Karin, M. 1999. How NF-kappaB is activated: the role of the IkappaB kinase (IKK) complex. *Oncogene* 18:6867-6874.

733. Zandi, E., and M. Karin. 1999. Bridging the gap: composition, regulation, and physiological function of the IkappaB kinase complex. *Mol Cell Biol* 19:4547-4551.

734. Tanaka, K., T. Kawakami, K. Tateishi, H. Yashiroda, and T. Chiba. 2001. Control of IkappaBalpha proteolysis by the ubiquitin-proteasome pathway. *Biochimie* 83:351-356.

735. Yamaoka, S., G. Courtois, C. Bessia, S. T. Whiteside, R. Weil, F. Agou, H. E. Kirk, R. J. Kay, and A. Israel. 1998. Complementation cloning of NEMO, a component of the IkappaB kinase complex essential for NF-kappaB activation. *Cell* 93:1231-1240.

736. Nakano, H., M. Shindo, S. Sakon, S. Nishinaka, M. Mihara, H. Yagita, and K. Okumura. 1998. Differential regulation of IkappaB kinase alpha and beta by two upstream kinases, NF-kappaB-inducing kinase and mitogen-activated protein kinase/ERK kinase kinase-1. *Proc Natl Acad Sci U S A* 95:3537-3542.

737. Zhong, H., R. E. Voll, and S. Ghosh. 1998. Phosphorylation of NF-kappa B p65 by PKA stimulates transcriptional activity by promoting a novel bivalent interaction with the coactivator CBP/p300. *Mol Cell* 1:661-671.

738. Schmitz, M. L., S. Bacher, and M. Kracht. 2001. I kappa B-independent control of NF-kappa B activity by modulatory phosphorylations. *Trends Biochem Sci* 26:186-190.

739. Sakurai, H., H. Chiba, H. Miyoshi, T. Sugita, and W. Toriumi. 1999. IkappaB kinases phosphorylate NF-kappaB p65 subunit on serine 536 in the transactivation domain. *J Biol Chem* 274:30353-30356.

740. Chinenov, Y., and T. K. Kerppola. 2001. Close encounters of many kinds: Fos-Jun interactions that mediate transcription regulatory specificity. *Oncogene* 20:2438-2452.

741. van Dam, H., and M. Castellazzi. 2001. Distinct roles of Jun : Fos and Jun : ATF dimers in oncogenesis. *Oncogene* 20:2453-2464.

742. Meyer-ter-Vehn, T., A. Covacci, M. Kist, and H. L. Pahl. 2000. Helicobacter pylori activates mitogen-activated protein kinase cascades and induces expression of the proto-oncogenes c-fos and c-jun. *J Biol Chem* 275:16064-16072.

743. Premkumar, D. R., G. Adhikary, J. L. Overholt, M. S. Simonson, N. S. Cherniack, and N. R. Prabhakar. 2000. Intracellular pathways linking hypoxia to activation of c-fos and AP-1. *Adv Exp Med Biol* 475:101-109.

744. Angel, P., K. Hattori, T. Smeal, and M. Karin. 1988. The jun proto-oncogene is positively autoregulated by its product, Jun/AP-1. *Cell* 55:875-885.

745. Binetruy, B., T. Smeal, and M. Karin. 1991. Ha-Ras augments c-Jun activity and stimulates phosphorylation of its activation domain. *Nature* 351:122-127.

746. Chou, S. Y., V. Baichwal, and J. E. Ferrell, Jr. 1992. Inhibition of c-Jun DNA binding by mitogen-activated protein kinase. *Mol Biol Cell* 3:1117-1130.

747. Shaw, G., and R. Kamen. 1986. A conserved AU sequence from the 3' untranslated region of GM-CSF mRNA mediates selective mRNA degradation. *Cell* 46:659-667.

748. Cooper, J. A., Jr., J. M. Fuller, K. M. McMinn, and R. R. Culbreth. 1998. Modulation of monocyte chemotactic protein-1 production by hyperoxia: importance of RNA stability in control of cytokine production. *Am J Respir Cell Mol Biol* 18:521-525.

749. Mukaida, N., S. Okamoto, Y. Ishikawa, and K. Matsushima. 1994. Molecular mechanism of interleukin-8 gene expression. *J Leukoc Biol* 56:554-558.

750. Poon, M., B. Liu, and M. B. Taubman. 1999. Identification of a novel dexamethasone-sensitive RNA-destabilizing region on rat monocyte chemoattractant protein 1 mRNA. *Mol Cell Biol* 19:6471-6478.

751. Shi, M. M., I. Chong, J. J. Godleski, and J. D. Paulauskis. 1999. Regulation of macrophage inflammatory protein-2 gene expression by oxidative stress in rat alveolar macrophages. *Immunology* 97:309-315.

752. Shi, M. M., J. J. Godleski, and J. D. Paulauskis. 1995. Molecular cloning and posttranscriptional regulation of macrophage inflammatory protein-1 alpha in alveolar macrophages. *Biochem Biophys Res Commun* 211:289-295.

753. Tebo, J. M., S. Datta, R. Kishore, M. Kolosov, J. A. Major, Y. Ohmori, and T. A. Hamilton. 2000. Interleukin-1-mediated stabilization of mouse KC mRNA depends on sequences in both 5'- and 3'-untranslated regions. *J Biol Chem* 275:12987-12993.

754. Jijon, H. B., W. J. Panenka, K. L. Madsen, and H. G. Parsons. 2002. MAP kinases contribute to IL-8 secretion by intestinal epithelial cells via a posttranscriptional mechanism. *Am J Physiol Cell Physiol* 283:C31-41.

755. Xiao, Y. Q., K. Someya, H. Morita, K. Takahashi, and K. Ohuchi. 1999. Involvement of p38 MAPK and ERK/MAPK pathways in staurosporine-induced production of macrophage inflammatory protein-2 in rat peritoneal neutrophils. *Biochim Biophys Acta* 1450:155-163.

756. Brennan, C. M., I. E. Gallouzi, and J. A. Steitz. 2000. Protein ligands to HuR modulate its interaction with target mRNAs in vivo. *J Cell Biol* 151:1-14.

757. Peng, S. S., C. Y. Chen, N. Xu, and A. B. Shyu. 1998. RNA stabilization by the AU-rich element binding protein, HuR, an ELAV protein. *EMBO J* 17:3461-3470.

758. Atasoy, U., S. L. Curry, I. Lopez de Silanes, A. B. Shyu, V. Casolaro, M. Gorospe, and C. Stellato. 2003. Regulation of eotaxin gene expression by TNF-alpha and IL-4 through mRNA stabilization: involvement of the RNA-binding protein HuR. *J Immunol* 171:4369-4378.

759. Fan, X. C., and J. A. Steitz. 1998. Overexpression of HuR, a nuclear-cytoplasmic shuttling protein, increases the in vivo stability of ARE-containing mRNAs. *EMBO J* 17:3448-3460.

760. Malter, J. S., and Y. Hong. 1991. A redox switch and phosphorylation are involved in the post-translational up-regulation of the adenosine-uridine binding factor by phorbol ester and ionophore. *J Biol Chem* 266:3167-3171.

761. Holtmann, H., J. Enninga, S. Kalble, A. Thiefes, A. Dorrie, M. Broemer, R. Winzen, A. Wilhelm, J. Ninomiya-Tsuji, K. Matsumoto, K. Resch, and M. Kracht. 2001. The MAPK kinase kinase TAK1 plays a central role in coupling the interleukin-1 receptor to both transcriptional and RNA-targeted mechanisms of gene regulation. *J Biol Chem* 276:3508-3516.

762. Goppelt-Struebe, M., and M. Stroebel. 1995. Synergistic induction of monocyte chemoattractant protein-1 (MCP-1) by platelet-derived growth factor and interleukin-1. *FEBS Lett* 374:375-378.

763. Leland Booth, J., and J. P. Metcalf. 1999. Type-specific induction of interleukin-8 by adenovirus. *Am J Respir Cell Mol Biol* 21:521-527.

764. Dai, Y., S. Datta, M. Novotny, and T. A. Hamilton. 2003. TGFbeta inhibits LPS-induced chemokine mRNA stabilization. *Blood* 102:1178-1185.

765. Wang, S. W., J. Pawlowski, S. T. Wathen, S. D. Kinney, H. S. Lichenstein, and C. L. Manthey. 1999. Cytokine mRNA decay is accelerated by an inhibitor of p38-mitogen-activated protein kinase. *Inflamm Res* 48:533-538.

766. Koga, T., E. Sardina, R. M. Tidwell, M. Pelletier, D. C. Look, and M. J. Holtzman. 1999. Virus-inducible expression of a host chemokine gene relies on replication-linked mRNA stabilization. *Proc Natl Acad Sci U S A* 96:5680-5685.

767. Yu, Y., H. Zeng, S. Lyons, A. Carlson, D. Merlin, A. S. Neish, and A. T. Gewirtz. 2003. TLR5-mediated activation of p38 MAPK regulates epithelial IL-8 expression via posttranscriptional mechanism. *Am J Physiol Gastrointest Liver Physiol* 285:G282-290.

768. Josse, C., J. R. Boelaert, M. Best-Belpomme, and J. Piette. 2001. Importance of post-transcriptional regulation of chemokine genes by oxidative stress. *Biochem J* 360:321-333.

769. Shi, M. M., J. J. Godleski, and J. D. Paulauskis. 1996. Regulation of macrophage inflammatory protein-1alpha mRNA by oxidative stress. *J Biol Chem* 271:5878-5883.

770. Page, S. M., G. J. Gleich, K. A. Roebuck, and L. L. Thomas. 1999. Stimulation of neutrophil interleukin-8 production by eosinophil granule major basic protein. *Am J Respir Cell Mol Biol* 21:230-237.

771. Chong, I. W., S. R. Lin, J. J. Hwang, M. S. Huang, T. H. Wang, J. Y. Hung, and J. D. Paulauskis. 2002. Expression and regulation of the macrophage inflammatory protein-1 alpha gene by nicotine in rat alveolar macrophages. *Eur Cytokine Netw* 13:242-249.

772. Chong, I. W., S. R. Lin, J. J. Hwang, M. S. Huang, T. H. Wang, M. S. Tsai, J. J. Hou, and J. D. Paulauskis. 2000. Expression and regulation of macrophage inflammatory protein-2 gene by vanadium in mouse macrophages. *Inflammation* 24:127-139.

773. Celada, A., A. Cruchaud, and L. H. Perrin. 1983. Assessment of immune phagocytosis of Plasmodium falciparum infected red blood cells by human monocytes and polymorphonuclear leukocytes. A method for visualizing infected red blood cells ingested by phagocytes. *J Immunol Methods* 63:263-271.

774. Jaramillo, M., D. C. Gowda, D. Radzioch, and M. Olivier. 2003. Hemozoin increases IFN-gamma-inducible macrophage nitric oxide generation through extracellular signal-regulated kinase- and NF-kappa B-dependent pathways. *J Immunol* 171:4243-4253.

775. Jaramillo, M., I. Plante, N. Ouellet, K. Vandal, P. A. Tessier, and M. Olivier. 2004. Hemozoin-inducible proinflammatory events in vivo: potential role in malaria infection. *J Immunol* 172:3101-3110.

776. Jaramillo, M., M. Godbout, and M. Olivier. 2005. Hemozoin induces macrophage chemokine expression through oxidative stress-dependent and -independent mechanisms. *J Immunol* 174:475-484.

777. Jaramillo, M., and M. Olivier. 2002. Hydrogen peroxide induces murine macrophage chemokine gene transcription via extracellular signal-regulated kinase- and cyclic adenosine 5'-monophosphate (cAMP)-dependent pathways: involvement of NF-kappa B, activator protein 1, and cAMP response element binding protein. *J Immunol* 169:7026-7038.

778. Hommel, M. 1993. Amplification of cytoadherence in cerebral malaria: towards a more rational explanation of disease pathophysiology. *Ann Trop Med Parasitol* 87:627-635.

779. Playfair, J. H., J. Taverne, C. A. Bate, and J. B. de Souza. 1990. The malaria vaccine: anti-parasite or anti-disease? *Immunol Today* 11:25-27.

780. Kremsner, P. G., S. Winkler, E. Wildling, J. Prada, U. Bienzle, W. Graninger, and A. K. Nussler. 1996. High plasma levels of nitrogen oxides are associated with severe disease and correlate with rapid parasitological and clinical cure in Plasmodium falciparum malaria. *Trans R Soc Trop Med Hyg* 90:44-47.

781. Nussler, A. K., W. Eling, and P. G. Kremsher. 1994. Patients with Plasmodium falciparum malaria and Plasmodium vivax malaria show increased nitrite and nitrate plasma levels. *J Infect Dis* 169:1418-1419.

782. Favre, N., B. Ryffel, G. Bordmann, and W. Rudin. 1997. The course of Plasmodium chabaudi chabaudi infections in interferon-gamma receptor deficient mice. *Parasite Immunol* 19:375-383.

783. Taylor-Robinson, A. W., R. S. Phillips, A. Severn, S. Moncada, and F. Y. Liew. 1993. The role of TH1 and TH2 cells in a rodent malaria infection. *Science* 260:1931-1934.

784. Kun, J. F., B. Mordmuller, D. J. Perkins, J. May, O. Mercereau-Puijalon, M. Alpers, J. B. Weinberg, and P. G. Kremsner. 2001. Nitric oxide synthase 2(Lambarene) (G-954C), increased nitric oxide production, and protection against malaria. *J Infect Dis* 184:330-336.

785. Stevenson, M. M., M. F. Tam, S. F. Wolf, and A. Sher. 1995. IL-12-induced protection against blood-stage Plasmodium chabaudi AS requires IFN-gamma and TNF-alpha and occurs via a nitric oxide-dependent mechanism. *J Immunol* 155:2545-2556.

786. Ahvazi, B. C., P. Jacobs, and M. M. Stevenson. 1995. Role of macrophage-derived nitric oxide in suppression of lymphocyte proliferation during blood-stage malaria. *J Leukoc Biol* 58:23-31.

787. Greenwood, B. M., A. M. Bradley-Moore, A. D. Bryceson, and A. Palit. 1972. Immunosuppression in children with malaria. *Lancet* 1:169-172.

788. Ockenhouse, C. F., S. Schulman, and H. L. Shear. 1984. Induction of crisis forms in the human malaria parasite Plasmodium falciparum by gamma-interferon-activated, monocyte-derived macrophages. *J Immunol* 133:1601-1608.

789. Held, T. K., X. Weihua, L. Yuan, D. V. Kalvakolanu, and A. S. Cross. 1999. Gamma interferon augments macrophage activation by lipopolysaccharide by two distinct mechanisms, at the signal transduction level and via an autocrine mechanism involving tumor necrosis factor alpha and interleukin-1. *Infect Immun* 67:206-212.

790. Drew, P. D., G. Franzoso, K. G. Becker, V. Bours, L. M. Carlson, U. Siebenlist, and K. Ozato. 1995. NF kappa B and interferon regulatory factor 1 physically interact and synergistically induce major histocompatibility class I gene expression. *J Interferon Cytokine Res* 15:1037-1045.

791. Jahnke, A., and J. P. Johnson. 1995. Intercellular adhesion molecule 1 (ICAM-1) is synergistically activated by TNF-alpha and IFN-gamma responsive sites. *Immunobiology* 193:305-314.

792. Neish, A. S., M. A. Read, D. Thanos, R. Pine, T. Maniatis, and T. Collins. 1995. Endothelial interferon regulatory factor 1 cooperates with NF-kappa B as a transcriptional activator of vascular cell adhesion molecule 1. *Mol Cell Biol* 15:2558-2569.

793. Ohmori, Y., and T. A. Hamilton. 1995. The interferon-stimulated response element and a kappa B site mediate synergistic induction of murine IP-10 gene transcription by IFN-gamma and TNF-alpha. *J Immunol* 154:5235-5244.

794. Bergeron, M., and M. Olivier. 2006. Trypanosoma cruzi-mediated IFN-gamma-inducible nitric oxide output in macrophages is regulated by iNOS mRNA stability. *J Immunol* 177:6271-6280.

795. Hamilton, T. A., and D. O. Adamson. 1987. Molecular mechanisms of signal transduction in macrophages. *Immunol. Today* 8:151.

796. Faure, V., C. Hecquet, Y. Courtois, and O. Goureau. 1999. Role of interferon regulatory factor-1 and mitogen-activated protein kinase pathways in the induction of nitric oxide synthase-2 in retinal pigmented epithelial cells. *J Biol Chem* 274:4794-4800.

797. Nakajima, T., S. Kinoshita, T. Sasagawa, K. Sasaki, M. Naruto, T. Kishimoto, and S. Akira. 1993. Phosphorylation at threonine-235 by a ras-dependent mitogen-activated protein kinase cascade is essential for transcription factor NF-IL6. *Proc Natl Acad Sci U S A* 90:2207-2211.

798. Screpanti, I., L. Romani, P. Musiani, A. Modesti, E. Fattori, D. Lazzaro, C. Sellitto, S. Scarpa, D. Bellavia, G. Lattanzio, and et al. 1995. Lymphoproliferative disorder and imbalanced T-helper response in C/EBP beta-deficient mice. *EMBO J* 14:1932-1941.

799. Su, B., and M. Karin. 1996. Mitogen-activated protein kinase cascades and regulation of gene expression. *Curr Opin Immunol* 8:402-411.

800. Forget, G., K. A. Siminovitch, S. Brochu, S. Rivest, D. Radzioch, and M. Olivier. 2001. Role of host phosphotyrosine phosphatase SHP-1 in the development of murine leishmaniasis. *Eur J Immunol* 31:3185-3196.

801. Pause, A., G. J. Belsham, A. C. Gingras, O. Donze, T. A. Lin, J. C. Lawrence, Jr., and N. Sonenberg. 1994. Insulin-dependent stimulation of protein synthesis by phosphorylation of a regulator of 5'-cap function. *Nature* 371:762-767.

802. Lin, T. A., X. Kong, T. A. Haystead, A. Pause, G. Belsham, N. Sonenberg, and J. C. Lawrence, Jr. 1994. PHAS-I as a link between mitogen-activated protein kinase and translation initiation. *Science* 266:653-656.

803. Caivano, M. 1998. Role of MAP kinase cascades in inducing arginine transporters and nitric oxide synthetase in RAW264 macrophages. *FEBS Lett* 429:249-253.

804. Hammermann, R., M. D. Dreissig, J. Mossner, M. Fuhrmann, L. Berrino, M. Gothert, and K. Racke. 2000. Nuclear factor-kappaB mediates simultaneous induction of inducible nitric-oxide synthase and Up-regulation of the cationic amino acid transporter CAT-2B in rat alveolar macrophages. *Mol Pharmacol* 58:1294-1302.

805. Nahrevanian, H., and M. J. Dascombe. 2001. Nitric oxide and reactive nitrogen intermediates during lethal and nonlethal strains of murine malaria. *Parasite Immunol* 23:491-501.

806. Taylor-Robinson, A. W. 1998. Immunoregulation of malarial infection: balancing the vices and virtues. *Int J Parasitol* 28:135-148.

807. Bancroft, C. C., Z. Chen, G. Dong, J. B. Sunwoo, N. Yeh, C. Park, and C. Van Waes. 2001. Coexpression of proangiogenic factors IL-8 and VEGF by human head and neck squamous cell carcinoma involves coactivation by MEK-MAPK and IKK-NF-kappaB signal pathways. *Clin Cancer Res* 7:435-442.

808. Kan, H., Z. Xie, and M. S. Finkel. 1999. TNF-alpha enhances cardiac myocyte NO production through MAP kinase-mediated NF-kappaB activation. *Am J Physiol* 277:H1641-1646.

809. Kim, K. W., S. H. Kim, E. Y. Lee, N. D. Kim, H. S. Kang, H. D. Kim, B. S. Chung, and C. D. Kang. 2001. Extracellular signal-regulated kinase/90-KDA ribosomal S6 kinase/nuclear factor-kappa B pathway mediates phorbol 12-myristate 13-acetate-induced megakaryocytic differentiation of K562 cells. *J Biol Chem* 276:13186-13191.

810. Yang, X., Y. Chen, and D. Gabuzda. 1999. ERK MAP kinase links cytokine signals to activation of latent HIV-1 infection by stimulating a cooperative interaction of AP-1 and NF-kappaB. *J Biol Chem* 274:27981-27988.

811. Lee, F. S., R. T. Peters, L. C. Dang, and T. Maniatis. 1998. MEKK1 activates both IkappaB kinase alpha and IkappaB kinase beta. *Proc Natl Acad Sci U S A* 95:9319-9324.

812. Barbeau, B., R. Bernier, N. Dumais, G. Briand, M. Olivier, R. Faure, B. I. Posner, and M. Tremblay. 1997. Activation of HIV-1 long terminal repeat transcription and virus replication via NF-kappaB-dependent and -independent pathways by potent phosphotyrosine phosphatase inhibitors, the peroxovanadium compounds. *J Biol Chem* 272:12968-12977.

813. Kang, J. L., I. S. Pack, S. M. Hong, H. S. Lee, and V. Castranova. 2000. Silica induces nuclear factor-kappa B activation through tyrosine phosphorylation of I kappa B-alpha in RAW264.7 macrophages. *Toxicol Appl Pharmacol* 169:59-65.

814. Au-Yeung, K. K., C. W. Woo, F. L. Sung, J. C. Yip, Y. L. Siow, and K. O. 2004. Hyperhomocysteinemia activates nuclear factor-kappaB in endothelial cells via oxidative stress. *Circ Res* 94:28-36.

815. Posner, B. I., R. Faure, J. W. Burgess, A. P. Bevan, D. Lachance, G. Zhang-Sun, I. G. Fantus, J. B. Ng, D. A. Hall, B. S. Lum, and et al. 1994. Peroxovanadium compounds. A new class of potent phosphotyrosine phosphatase inhibitors which are insulin mimetics. *J Biol Chem* 269:4596-4604.

816. Olivier, M., B. J. Romero-Gallo, C. Matte, J. Blanchette, B. I. Posner, M. J. Tremblay, and R. Faure. 1998. Modulation of interferon-gamma-induced macrophage activation by phosphotyrosine phosphatases inhibition. Effect on murine Leishmaniasis progression. *J Biol Chem* 273:13944-13949.

817. Forget, G., C. Matte, K. A. Siminovitch, S. Rivest, P. Pouliot, and M. Olivier. 2005. Regulation of the Leishmania-induced innate inflammatory response by the protein tyrosine phosphatase SHP-1. *Eur J Immunol* 35:1906-1917.

818. Beraud, C., W. J. Henzel, and P. A. Baeuerle. 1999. Involvement of regulatory and catalytic subunits of phosphoinositide 3-kinase in NF-kappaB activation. *Proc Natl Acad Sci U S A* 96:429-434.

819. Fan, C., Q. Li, D. Ross, and J. F. Engelhardt. 2003. Tyrosine phosphorylation of I kappa B alpha activates NF kappa B through a redox-regulated and c-Src-dependent mechanism following hypoxia/reoxygenation. *J Biol Chem* 278:2072-2080.

820. Takada, Y., A. Mukhopadhyay, G. C. Kundu, G. H. Mahabeleshwar, S. Singh, and B. B. Aggarwal. 2003. Hydrogen peroxide activates NF-kappa B through tyrosine phosphorylation of I kappa B alpha and serine phosphorylation of p65: evidence for the involvement of I kappa B alpha kinase and Syk protein-tyrosine kinase. *J Biol Chem* 278:24233-24241.

821. Lee, K., and W. J. Esselman. 2001. cAMP potentiates H(2)O(2)-induced ERK1/2 phosphorylation without the requirement for MEK1/2 phosphorylation. *Cell Signal* 13:645-652.

822. Bayatti, N., and J. Engele. 2001. Cyclic AMP modulates the response of central nervous system glia to fibroblast growth factor-2 by redirecting signalling pathways. *J Neurochem* 78:972-980.

823. Jornot, L., H. Petersen, and A. F. Junod. 1997. Modulation of the DNA binding activity of transcription factors CREP, NFkappaB and HSF by H2O2 and TNF alpha. Differences between in vivo and in vitro effects. *FEBS Lett* 416:381-386.

824. Guyton, K. Z., Y. Liu, M. Gorospe, Q. Xu, and N. J. Holbrook. 1996. Activation of mitogen-activated protein kinase by H2O2. Role in cell survival following oxidant injury. *J Biol Chem* 271:4138-4142.

825. Janssen-Heininger, Y. M., I. Macara, and B. T. Mossman. 1999. Cooperativity between oxidants and tumor necrosis factor in the activation of nuclear factor (NF)-kappaB: requirement of Ras/mitogen-activated protein kinases in the activation of NF-kappaB by oxidants. *Am J Respir Cell Mol Biol* 20:942-952.

826. Kumaratilake, L. M., A. Ferrante, and C. M. Rzepczyk. 1990. Tumor necrosis factor enhances neutrophil-mediated killing of Plasmodium falciparum. *Infect Immun* 58:788-793.

827. Nnalue, N. A., and M. J. Friedman. 1988. Evidence for a neutrophil-mediated protective response in malaria. *Parasite Immunol* 10:47-58.

828. Golenser, J., M. Kamyl, A. Tsafack, E. Marva, A. Cohen, N. Kitrossky, and M. Chevion. 1992. Correlation between destruction of malarial parasites by polymorphonuclear leucocytes and oxidative stress. *Free Radic Res Commun* 17:249-262.

829. Salmon, D., J. L. Vilde, B. Andrieu, R. Simonovic, and J. Lebras. 1986. Role of immune serum and complement in stimulation of the metabolic burst of human neutrophils by Plasmodium falciparum. *Infect Immun* 51:801-806.

830. Pukrittayakamee, S., R. Clemens, C. Pramoolsinsap, H. E. Karges, S. Vanijanonta, D. Bunnag, and N. J. White. 1992. Polymorphonuclear leucocyte elastase in Plasmodium falciparum malaria. *Trans R Soc Trop Med Hyg* 86:598-601.

831. Dockrell, H. M., A. Alavi, and J. H. Playfair. 1986. Changes in oxidative burst capacity during murine malaria and the effect of vaccination. *Clin Exp Immunol* 66:37-43.

832. Bautista, A. P. 2001. Free radicals, chemokines, and cell injury in HIV-1 and SIV infections and alcoholic hepatitis. *Free Radic Biol Med* 31:1527-1532.

833. Rodriguez-Acosta, A., H. J. Finol, M. Pulido-Mendez, A. Marquez, G. Andrade, N. Gonzalez, I. Aguilar, M. E. Giron, and A. Pinto. 1998. Liver ultrastructural pathology in mice infected with Plasmodium berghei. *J Submicrosc Cytol Pathol* 30:299-307.

834. Descamps-Latscha, B., F. Lunel-Fabiani, A. Kara-Binis, and P. Druilhe. 1987. Generation of reactive oxygen species in whole blood from patients with acute falciparum malaria. *Parasite Immunol* 9:275-279.

835. Matzuk, M. M., L. Dionne, Q. Guo, T. R. Kumar, and R. M. Lebovitz. 1998. Ovarian function in superoxide dismutase 1 and 2 knockout mice. *Endocrinology* 139:4008-4011.

836. Eugui, E. M., and A. C. Allison. 1980. Differences in susceptibility of various mouse strains to haemoprotozoan infections: possible correlation with natural killer activity. *Parasite Immunol* 2:277-292.

837. Allison, A. C., and E. M. Eugui. 1983. The role of cell-mediated immune responses in resistance to malaria, with special reference to oxidant stress. *Annu Rev Immunol* 1:361-392.

838. Calkins, C. M., D. D. Bensard, B. D. Shames, E. J. Pulido, E. Abraham, N. Fernandez, X. Meng, C. A. Dinarello, and R. C. McIntyre, Jr. 2002. IL-1 regulates in vivo C-X-C chemokine induction and neutrophil sequestration following endotoxemia. *J Endotoxin Res* 8:59-67.

839. Romano, M., M. Sironi, C. Toniatti, N. Polentarutti, P. Fruscella, P. Ghezzi, R. Faggioni, W. Luini, V. van Hinsbergh, S. Sozzani, F. Bussolino, V. Poli, G. Ciliberto, and A. Mantovani. 1997. Role of IL-6 and its soluble receptor in induction of chemokines and leukocyte recruitment. *Immunity* 6:315-325.

840. Larsen, C. M., K. A. Wadt, L. F. Juhl, H. U. Andersen, A. E. Karlsen, M. S. Su, K. Seedorf, L. Shapiro, C. A. Dinarello, and T. Mandrup-Poulsen. 1998. Interleukin-1beta-induced rat pancreatic islet nitric oxide synthesis requires both the p38 and extracellular signal-regulated kinase 1/2 mitogen-activated protein kinases. *J Biol Chem* 273:15294-15300.

841. Grau, G. E., T. E. Taylor, M. E. Molyneux, J. J. Wirima, P. Vassalli, M. Hommel, and P. H. Lambert. 1989. Tumor necrosis factor and disease severity in children with falciparum malaria. *N Engl J Med* 320:1586-1591.

842. Clark, I. A., S. Ilschner, J. D. MacMicking, and W. B. Cowden. 1990. TNF and Plasmodium berghei ANKA-induced cerebral malaria. *Immunol Lett* 25:195-198.

843. Pavia, C. S. 1983. Expression of cell-mediated antimicrobial immunity by mouse trophoblast monolayers. *J Infect Dis* 147:1006-1010.

844. Chandramohan, D., and B. M. Greenwood. 1998. Is there an interaction between human immunodeficiency virus and Plasmodium falciparum? *Int J Epidemiol* 27:296-301.

845. Salvatori, F., and G. Scarlatti. 2001. HIV type 1 chemokine receptor usage in mother-to-child transmission. *AIDS Res Hum Retroviruses* 17:925-935.

846. Doranz, B. J., J. Rucker, Y. Yi, R. J. Smyth, M. Samson, S. C. Peiper, M. Parmentier, R. G. Collman, and R. W. Doms. 1996. A dual-tropic primary HIV-1 isolate that uses fusin and the beta-chemokine receptors CKR-5, CKR-3, and CKR-2b as fusion cofactors. *Cell* 85:1149-1158.

847. Feng, Y., C. C. Broder, P. E. Kennedy, and E. A. Berger. 1996. HIV-1 entry cofactor: functional cDNA cloning of a seven-transmembrane, G protein-coupled receptor. *Science* 272:872-877.

848. Ma, N., N. H. Hunt, M. C. Madigan, and T. Chan-Ling. 1996. Correlation between enhanced vascular permeability, up-regulation of cellular adhesion molecules and monocyte adhesion to the endothelium in the retina during the development of fatal murine cerebral malaria. *Am J Pathol* 149:1745-1762.

849. Rest, J. R. 1982. Cerebral malaria in inbred mice. I. A new model and its pathology. *Trans R Soc Trop Med Hyg* 76:410-415.

850. Newton, R. A., and N. Hogg. 1998. The human S100 protein MRP-14 is a novel activator of the beta 2 integrin Mac-1 on neutrophils. *J Immunol* 160:1427-1435.

851. Ryckman, C., S. R. McColl, K. Vandal, R. de Medicis, A. Lussier, P. E. Poubelle, and P. A. Tessier. 2003. Role of S100A8 and S100A9 in neutrophil recruitment in response to monosodium urate monohydrate crystals in the air-pouch model of acute gouty arthritis. *Arthritis Rheum* 48:2310-2320.

852. Belnoue, E., F. T. Costa, A. M. Vigario, T. Voza, F. Gonnet, I. Landau, N. Van Rooijen, M. Mack, W. A. Kuziel, and L. Renia. 2003. Chemokine receptor CCR2 is not essential for the development of experimental cerebral malaria. *Infect Immun* 71:3648-3651.

853. Kumar, K. A., Y. Rajgopal, U. Pillai, and P. P. Babu. 2003. Activation of nuclear transcription factor-kappa B is associated with the induction of inhibitory kappa B kinase-beta and involves differential activation of protein kinase C and protein tyrosine kinases during fatal murine cerebral malaria. *Neurosci Lett* 340:139-142.

854. Chan, P. H., and R. A. Fishman. 1984. The role of arachidonic acid in vasogenic brain edema. *Fed Proc* 43:210-213.

855. Boutlis, C. S., D. C. Gowda, R. S. Naik, G. P. Maguire, C. S. Mgone, M. J. Bockarie, M. Lagog, E. Ibam, K. Lorry, and N. M. Anstey. 2002. Antibodies to Plasmodium falciparum glycosylphosphatidylinositols: inverse association with tolerance of parasitemia in Papua New Guinean children and adults. *Infect Immun* 70:5052-5057.

856. de Souza, J. B., J. Todd, G. Krishegowda, D. C. Gowda, D. Kwiatkowski, and E. M. Riley. 2002. Prevalence and boosting of antibodies to Plasmodium falciparum glycosylphosphatidylinositols and evaluation of their association with protection from mild and severe clinical malaria. *Infect Immun* 70:5045-5051.

843645

Printed in Great Britain by
Amazon.co.uk, Ltd.,
Marston Gate.